ENCYCLOPEDIA OF
ANIMALS

A FAMILY REFERENCE GUIDE

First published by Parragon in 2012

Parragon
Queen Street House
4 Queen Street
Bath BA1 1HE, UK
www.parragon.com

Designed by Design Principals

ISBN 978-1-4454-8470-9
Printed in China

ENCYCLOPEDIA OF
ANIMALS

A FAMILY REFERENCE GUIDE

PaRragon

Bath•New York•Singapore•Hong Kong•Cologne•Delhi
Melbourne•Amsterdam•Johannesburg•Shenzhen

CONTENTS

AMAZING ANIMALS

 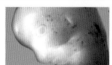    

Welcome to the amazing world of animals! In this fascinating book, with its wealth of wonderful photos and intriguing facts, you can explore the astonishing lives of the millions of species with which we share the planet earth.

From the frozen poles to the sunless depths of oceans, from highest mountains to arid deserts, every region of our planet has been colonized by animals. Animals live in interactive partnerships with other species and with their physical environments. Without animals, humans could not survive. As scientific understanding of the animal world increases, so does our appreciation of the intimate relationship between humans and animals, and of the need to conserve the rich diversity of life on earth.

The *Family Encyclopedia of Animals* looks at five main groups of animals: Mammals, Birds, Reptiles, Fish and Amphibians, and Invertebrates. Mammals are the smallest group, but the mammals include the largest animal, the blue whale. Easily the biggest group, with myriads of individual animals in millions of species, are the Invertebrates, the group that includes insects as well as crustaceans and arachnids.

Mammals

From the frozen tundra polar ice to the tropical rain forests, the earth is rich in mammals, animals of astonishing diversity that have occupied a range of environments. Mammals probably began to dominate the earth about 65 million years ago. Without a doubt, modern humans are the most successful mammals— we have explored and colonized all the earth's habitats! Our domestic coexistence with other species began around 10,000 years ago, when human culture shifted from nomadic hunter-gatherers to a society based on agriculture. Humans began to domesticate some mammals for work, meat and milk, and for useful products, such as wool and leather. These domestic animals included dogs, sheep, pigs, cows, goats, and horses.

There are 5,416 known mammal species. Their diversity is such that the smallest mammal, the shrew,

COMMUNICATION
The ways in which cetaceans communicate with others of their kind are among the most sophisticated in the animal kingdom. Dolphins, for example, click with their mandibles when in trouble and whistle repeatedly when afraid or excited.

may weigh only $\frac{1}{10}$ oz (3 g), while the largest, the blue whale, can reach up to 160 tons (145 tonnes). But this diversity is also evident in mammals' adaptation to different environments. Mammal species can run, glide, fly, jump, swim, and crawl. To survive the rigors of low temperatures, some cold-climate animals—such as pregnant female polar bears—spend the winter in deep sleep to save energy. Most aquatic mammals have thick layers of body fat instead of hair, which in most land mammals serves to conserve heat. Seals, dolphins, bats, and chimpanzees all have upper limbs with similar bones, but seals have flippers, dolphins have fins, bats have wings, and chimpanzees have arms.

Birds
Birds never cease to amaze us. Avian abilities are varied, including diving, swimming, and complex nest-building skills, but their ability to fly has long been envied by humans. Equally astonishing is the distance some birds fly. It has been estimated that more than 200 million birds migrate each year all over the planet. Many birds fly thousands of miles, crossing deserts and oceans to reach their breeding and feeding grounds. Much of bird behavior, such as the ability of migrants to find their way across continents and back, continues to be a mystery.

It is believed that there are approximately 9,700 bird species in the world. This makes birds the second largest vertebrate group of animals, after fish. Birds vary greatly in size: from a hummingbird weighing just $\frac{3}{50}$ ounce (1.6 g), to an ostrich weighing in at 330 pounds (150 kg). Although most birds fly, there are some flightless birds—such as kiwis, penguins, rheas, and ostriches. Other birds are adapted to aquatic life in oceans, rivers, and lakes. The shape of birds' feet and bills reflects adaptations. Some aquatic species have bills modified to filter feed on small particles in water, whereas birds of prey use hooked bills to hold down and tear apart prey. In many bird species, both males and females share nest making and the rearing of young, and some birds also display social behaviors in groups.

Reptiles

There are about 8,200 species of reptiles. They include turtles, lizards, snakes, crocodiles, and tuataras. Reptiles were the first vertebrate animals to be independent of water. An amniotic egg with a waterproof shell enabled reptiles to breed and hatch on land, without the need to return to water. Because reptiles rely on external heat to regulate their body temperature, many reptiles spend hours in the sun warming themselves by infrared radiation.

Humans have long feared and respected reptiles. Snakes, crocodiles, and mythical dragons are found in the legends of peoples throughout the world, and these animals are often associated with magic. Many reptile species have impressive abilities: they can scale walls, burrow, swim, climb slender stalks, and even run across loose, hot sand dunes. Chameleons have stereoscopic eyes that allow them to see prey with three-dimensional depth perception. While snakes have no visible ears, they are able to hear through their jaw and internal ear bones. They can detect low-frequency vibrations in the soil to spot approaching predators or prey. Some snakes can swallow prey larger than themselves.

PARTRIDGE EGGS
The female lays eggs at intervals of one to two days, and she is the one who incubates them.

KOMODO DRAGON
These endangered lizards live only on a group of islands in Indonesia. Their saliva is full of bacteria that can kill their prey with only one bite. They can detect other Komodo dragons from several miles away.

Although snakes have been hated and persecuted, in fact, only one out of ten species of snake is dangerous. Most snakes will never attack unless they feel threatened and use warning mechanisms and behaviors before attacking. Learning which species may be dangerous is important to protect snakes and people living alongside these animals. Many reptile species today are in danger of extinction, threatened by hunting and habitat destruction. These amazing animals with extraordinary traits have been around for many millions of years.

Fish and Amphibians

Fish and amphibians were among the first vertebrates, creatures with internal skeletons, and each species has evolved to help it survive in a specific habitat. Fish are uniquely adapted to the watery world, with gills for breathing and fins for swimming. They can live in oceans, lakes, and freshwater rivers and streams. Some of the world's most exotic and brilliantly colored animals are the fish of the warm tropical oceans, while strange and little-known fish inhabit the cold, dark ocean deeps. Many fish are valued as food by people, and the conservation of commercial fish stocks is a key issue for the future. Some fish remain elusive and even feared; few animals match the reputation in fact and fiction of the shark.

Way back in evolutionary prehistory, some species moved from water to dry land, breathing by means of lunglike air sacs. Fish with proto-limbs, known as lobe-fin fishes, were able to exploit new food

GREEN MORAY
Unlike most fish, the moray has no scales. It excretes a slippery film to cover its thick, muscular body and protect itself from parasites. The moray hunts at night and detects its prey with its excellent sense of smell.

sources and, over time, adapted to life on land. This evolutionary change—from water-dwelling to land-dwelling animals—constituted a revolution for life on the earth. Some land-dwelling animals retained a link with water through breeding behavior; these are the amphibians, such as frogs and toads. The living amphibians are a tiny fraction of a once-numerous class that appeared during the Devonian period of prehistory, but most became extinct during the later Triassic period.

The Invertebrates
The most ancient forms of animal life on the earth are the invertebrates. They are also by far the most numerous. With more than 1.5 million known species, it is estimated that 97 percent of all animal species are invertebrates.

Invertebrates exist in a variety of habitats and forms. Some are soft-bodied animals, such as worms or jellyfish, while others, such as insects and crustaceans, have a hard external skeleton for protection. This fascinating group of creatures offers amazing examples of adaptation and habits. Mosquitoes can pierce the skin of mammals and feed on their blood. Flies can eat solid food because their digestive process begins outside their bodies. No human athlete can match the jumping prowess of the tiny wingless flea.

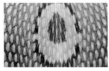
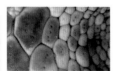

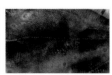
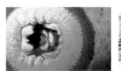
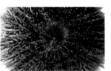

Many invertebrates play a vital role in food chains and ecosystems, so they are very important to other species. Bees and other pollinating insects are essential to the life cycle of many flowering plants that provide food for humans and other animals. Other invertebrates are useful predators and controls. Did you know that there are 35,000 known species of spiders (only 30 are dangerous to people), and that without spiders as expert predators, the world might be overrun by insect pests?

Explore the World of Animals

In this fact-filled book, with its many dazzling photographs and illustrations, we invite you to explore the amazing world of animals. Read about life cycles, social behavior, and special features and characteristics. Get close to the fascinating creatures that share our world. Learn how they are able to run, climb, swim, dive, and fly. Discover how they find food and raise their young.

We may wonder at the thousands of species of animals alive today. Yet it is thought that 99.9 percent of all the animal species that have ever existed are now extinct. Extinctions happen naturally, but human actions can also lead to species being lost forever. Modern science is showing us how all living things are interconnected in complex ecosystems. By learning more about animals, we will be better able to conserve the natural world and to protect the earth's wildlife for future generations.

PEARL FORMATION
Occasionally, grains of sand or parasites accidentally become lodged in an oyster's body and cannot escape. To relieve the annoyance, the oyster secretes a smooth, hard, crystalline substance called nacre around the object.

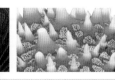

1 MAMMALS

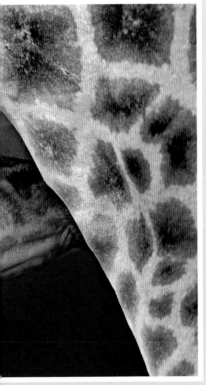

30

54

What They Are Like

All mammals have stereoscopic vision, which gives them depth perception. Moreover, in the case of hunters, such as tigers, their night vision is six times keener than that of humans. There are many species that have a very keen sense of smell, and the sense of taste is closely

BENGAL TIGER
Panthera tigris tigris is the largest member of the feline family, easily recognized by its orange fur with black stripes and white spots.

linked to that of smell. Hair, too, performs various functions in these animals' lives—conserving body heat, providing protection, and serving as camouflage.

Those that have almost no hair and live in environments where the temperature is very low, such as whales, have developed a layer of fat under their skins.

What Is a Mammal?

Mammals share a series of characteristics that distinguish their class: a body covered by hair, the birth of live young, and the feeding of newborns on milk produced by the females' mammary glands. All breathe through lungs, and all possess a closed, double circulatory system and the most developed nervous systems in the animal kingdom. The ability to maintain a constant body temperature has allowed them to spread out and conquer every corner of the earth, from the coldest climates to hot deserts and from the mountains to oceans.

GORILLA
Gorilla gorilla

A Body for Every Environment

Skin covered with hair and sweat glands helps create and maintain a constant body temperature. At the same time, with eyes placed on each side of the head (monocular vision, with the sole exception of the primates, which have binocular vision), they are afforded important angles of sight. Limbs are either of the foot or chiridium type, with slight variations depending on the part of the foot used for walking. In aquatic mammals, the limbs have evolved into fins; in bats, into wings. Hunters have powerful claws, and unguligrades (such as horses) have strong hooves that support the whole body when running.

5,416
THE NUMBER OF MAMMAL SPECIES ESTIMATED TO EXIST ON EARTH

BOTTLENOSE DOLPHIN
Tursiops truncatus

Hair

Body hair is unique to mammals and absent in other classes of animals. Sirenians, with little hair, and cetaceans are exceptions; in both cases, the absence of hair is a result of the mammal's adaptation to an aquatic environment.

Dentition

The majority of mammals change dentition in their passage to adulthood. Teeth are specialized for each function: molars for chewing, canines for tearing, and incisors for gnawing. In rodents, such as chipmunks, the teeth are renewed by continuous growth.

CHIPMUNK
Family sciuridae

Close Relatives

Humans belong to the primate group. Hominids (orangutans, gorillas, and chimpanzees) are the largest of these, weighing between 105 and 595 pounds (48–270 kg). In general, males are larger than females, with robust bodies and well-developed arms. Their vertical carriage differentiates their skeletons from those of other primates. Gorillas inhabit only the equatorial jungles of western Africa. They support themselves on their forelimbs while walking. Normally their height varies between 4 and 6 feet (1.2–1.8 m), but, if they raise their forelimbs and stand erect, they can be over 6½ feet tall (2 m).

CRANIUM
Relatively large compared to the size of the body. And the brain is more developed and more complex than that of any other animal.

ALWAYS 98°F (37°C)
The ability to maintain a constant body temperature is not a characteristic unique to mammals; birds also have that ability.

AN EAR OF BONES
The tiny bones of the ear form a system for sensing and transmitting sound.

LOWER JAW
Formed by a single bone, called the dentary, and teeth specialized for each function. The entire cranium has a very simplified bone structure.

MAMMARY GLANDS
Secrete the milk with which the females feed their young during their first months of life. These glands give the class its name.

A THICK SKIN
Formed by an outer layer (epidermis), another deeper layer (dermis), and a fatty substratum that contributes to homeothermy.

Homeothermy

The ability to keep body temperature relatively constant, independent of the ambient temperature. Hibernating species are the exception; they must lower their body temperature to enter into this state of reduced metabolic activity. Contrary to popular belief, bears do not truly hibernate but rather enter into a period of deep sleep during winter.

GRIZZLY BEAR (BROWN BEAR)
Ursus arctos

Limbs

Mammals have four limbs that are adapted for moving about on land. Their forelimbs have certain other abilities (swimming, manipulation, attack and defense, protection). The exceptions are the cetaceans, so adapted to marine life that they have only two fingerless limbs, and seals (phocidae).

ELEPHANT SEALS
Family phocidae

Take Habitat into Account

Between every mammal and its natural habitat there is a relationship that exists and is expressed in the animal's physical characteristics. Just as the flippers of the elephant seal are used to swim and hunt fish, mimicry and running are vital for deer. Physiology is a special instrument of adaptation to the environment, as in the case of the camel.

Aquatic	Temperate forests	Desert	Meadow or pastureland

Tropical savanna	Tropical rain forest	Taiga	Tundra

AN UNCOMMON PRIMATE

Humans have adapted to almost all habitats through their ability to modify certain elements of their habitat to their advantage. They often create tools to help them adapt to their environment. In this way, they do not need to rely on natural evolution alone. Humans have adapted to almost all habitats.

Constant Heat

Mammals are homeothermic—which means they are capable of maintaining a stable internal body temperature despite environmental conditions. This ability has allowed them to establish themselves in every region of the planet. Homeostasis is achieved by a series of processes that tend to keep water levels and concentrations of minerals and glucose in the blood in equilibrium as well as prevent an accumulation of waste products—among other things.

Great Swimmers

Polar bears swim with ease in open waters and reach a speed of 6 miles an hour (10 km/h). They propel themselves with their great front paws and use their back feet as rudders. The bear's hair is hollow and filled with air, which helps with buoyancy. When the bear dives, its eyes remain open.

Kings of the Arctic

The polar bear (or white bear) is a perfect example of adaptation to the inhospitable environment it inhabits. The fur, which can appear white, pale yellow, or cream in color, is actually translucent and colorless; it consists of two layers, one with thick short hairs and a superficial one with long hairs. The insulation necessary to survive in the Arctic is provided by its thick fur and by a layer of fat under the skin, both of which allow the polar bear to dive and swim in icy waters and to withstand blizzards.

POLAR BEAR
Ursus maritimus

SHELTERED CUBS

The cubs are born in winter, and the skin of the mother generates heat that protects the cubs from the extreme cold.

Migration

When spring begins, these bears travel south, escaping the break up of the arctic ice.

Metabolism

The layer of fat is between 4 and 6 inches (10–15 cm) thick and provides not only thermal insulation but also an energy reserve. When the temperature reaches critical levels—at the Pole it can drop to between -60° and -75°F (-50° to -60°C)—the animal's metabolism increases and begins to rapidly burn energy from fat and food. In this way, the polar bear maintains its body temperature.

UNDER THE ICE

Females dig a tunnel in the spring; when they become pregnant, they hibernate without eating and can lose 45 percent of their weight.

SECONDARY ACCESS TUNNEL

RESPIRATORY PATHWAYS
The bears have membranes in their snouts that warm and humidify the air before it reaches the lungs.

HAIR
An impermeable, translucent surface

Hollow chamber with air

CHAMBER OR REFUGE

MAIN ACCESS TUNNEL

ENTRANCE

LAYERS

GUARD HAIRS
Outer

UNDERFUR
Inner

FAT
4–6 inches (10–15 cm) thick

PRINCIPAL FAT RESERVES
Thighs, haunches, and abdomen

Curling Up

Many cold-climate mammals curl up into balls, covering their extremities and bending their tails over their bodies as a kind of blanket. In this way, the surface area subjected to heat loss will be minimal. Hot-climate animals stretch out their bodies to dissipate heat.

more than
6 miles (10 km)

PER HOUR IS THE AVERAGE SPEED AT WHICH POLAR BEARS SWIM.

SLOW AND STEADY SWIMMING

HIND LEGS
function as a rudder.

FORELIMBS
function as a motor.

AND FINALLY . . . THE FLOATING SLAB
When they tire of swimming, they rest, floating. They manage to cross distances of more than 37 miles (60 km) in this manner.

TO GET OUT: ANTISLIP PALMS
Their palms have surfaces with small papillae that create friction with ice, keeping them from slipping.

HYDRODYNAMIC ANATOMY

Grace and Movement

Horses, one of the odd-toed, hoofed, ungulate mammals, are considered symbols of grace and freedom. They have great vigor and can run swiftly because their spine bends very little, preventing unnecessary expenditure of energy during the rising and falling of their body mass. They are equipped with strong, light, and flexible bones, and their muscles work by contraction, arranged in pairs or groups that pull in opposing directions.

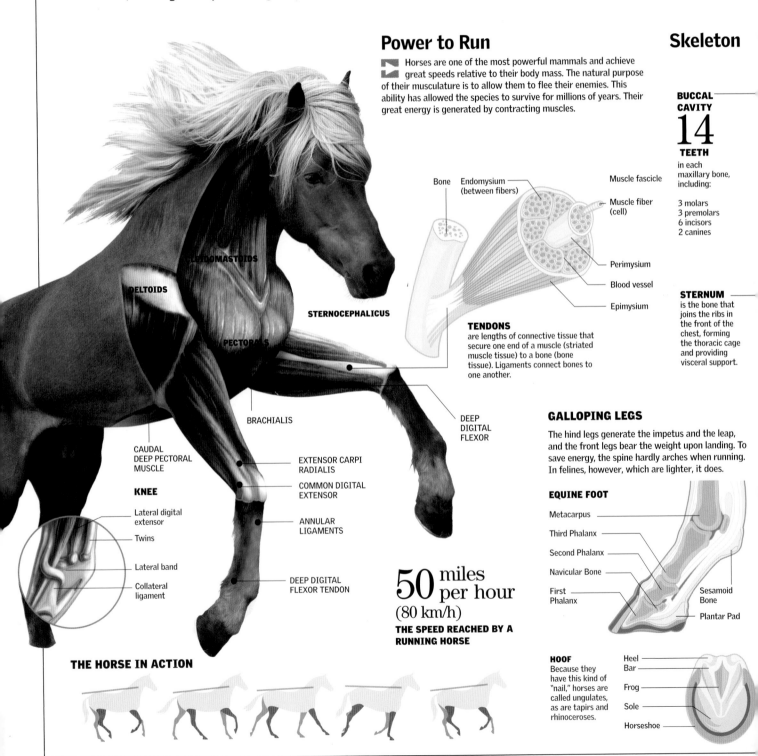

Power to Run

Horses are one of the most powerful mammals and achieve great speeds relative to their body mass. The natural purpose of their musculature is to allow them to flee their enemies. This ability has allowed the species to survive for millions of years. Their great energy is generated by contracting muscles.

Bone
Endomysium (between fibers)
Muscle fascicle
Muscle fiber (cell)
Perimysium
Blood vessel
Epimysium

TENDONS
are lengths of connective tissue that secure one end of a muscle (striated muscle tissue) to a bone (bone tissue). Ligaments connect bones to one another.

STERNOCLEIDOMASTOIDS
DELTOIDS
STERNOCEPHALICUS
PECTORALS

BRACHIALIS

DEEP DIGITAL FLEXOR

CAUDAL DEEP PECTORAL MUSCLE

EXTENSOR CARPI RADIALIS

COMMON DIGITAL EXTENSOR

KNEE

Lateral digital extensor
Twins
Lateral band
Collateral ligament

ANNULAR LIGAMENTS

DEEP DIGITAL FLEXOR TENDON

50 **miles per hour**
(80 km/h)
THE SPEED REACHED BY A RUNNING HORSE

Skeleton

BUCCAL CAVITY
14
TEETH
in each maxillary bone, including:

3 molars
3 premolars
6 incisors
2 canines

STERNUM
is the bone that joins the ribs in the front of the chest, forming the thoracic cage and providing visceral support.

GALLOPING LEGS

The hind legs generate the impetus and the leap, and the front legs bear the weight upon landing. To save energy, the spine hardly arches when running. In felines, however, which are lighter, it does.

EQUINE FOOT

Metacarpus
Third Phalanx
Second Phalanx
Navicular Bone
First Phalanx
Sesamoid Bone
Plantar Pad

HOOF
Because they have this kind of "nail," horses are called ungulates, as are tapirs and rhinoceroses.

Heel
Bar
Frog
Sole
Horseshoe

THE HORSE IN ACTION

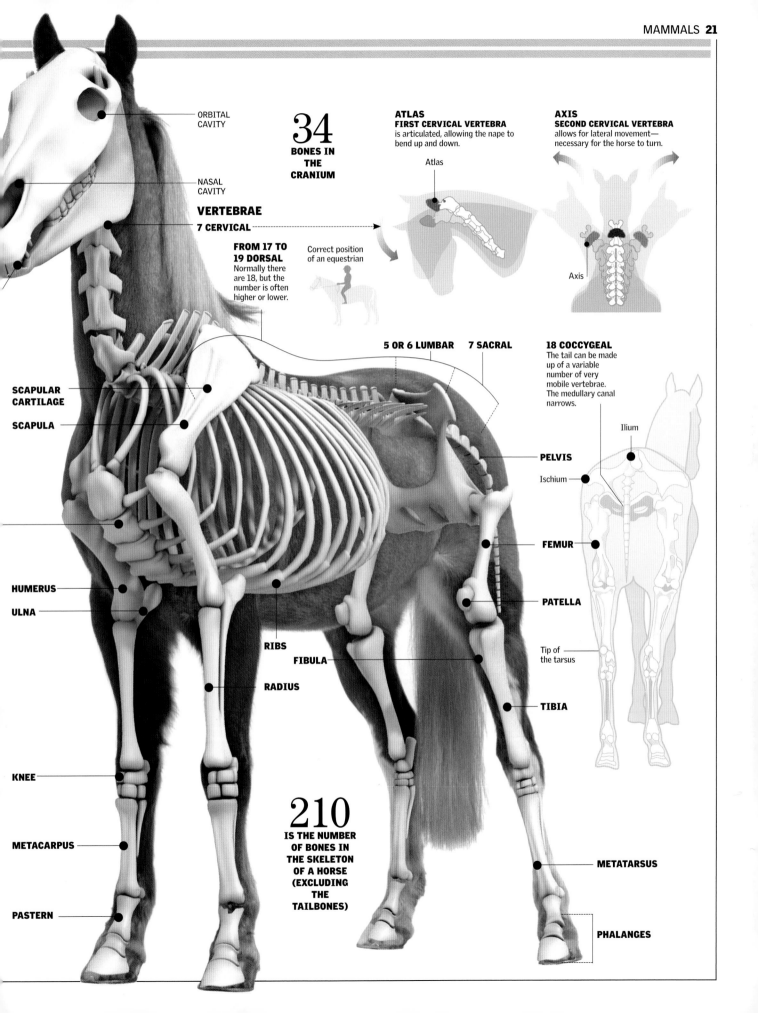

ORBITAL
CAVITY

NASAL
CAVITY

34
**BONES IN
THE
CRANIUM**

VERTEBRAE

7 CERVICAL

**FROM 17 TO
19 DORSAL**
Normally there
are 18, but the
number is often
higher or lower.

Correct position
of an equestrian

**ATLAS
FIRST CERVICAL VERTEBRA**
is articulated, allowing the nape to
bend up and down.

Atlas

**AXIS
SECOND CERVICAL VERTEBRA**
allows for lateral movement—
necessary for the horse to turn.

Axis

**SCAPULAR
CARTILAGE**

SCAPULA

5 OR 6 LUMBAR   **7 SACRAL**

18 COCCYGEAL
The tail can be made
up of a variable
number of very
mobile vertebrae.
The medullary canal
narrows.

Ilium

PELVIS

Ischium

FEMUR

HUMERUS

ULNA

PATELLA

RIBS

FIBULA

RADIUS

Tip of
the tarsus

TIBIA

KNEE

210
**IS THE NUMBER
OF BONES IN
THE SKELETON
OF A HORSE
(EXCLUDING
THE
TAILBONES)**

METACARPUS

METATARSUS

PASTERN

PHALANGES

Extremities

Mammals' extremities are basically either of the foot or chiridium type but are modified according to the way in which each species moves about. Thus, for example, they become fins for swimming in aquatic mammals and membranous wings in bats. In land mammals, these variations depend on the way the animal bears its weight in walking: those that use the whole foot are called plantigrades; those that place their weight on their digits, digitigrades; and those that only touch the ground with the tips of their phalanges, ungulates.

Functionally Adapted

Another criterion for classifying mammals by their legs, in addition to their morphology, is the function the legs perform. Cats, dogs, and horses have four limbs for locomotion. Primates have differentiated forelimbs, and they also use legs to capture food or bring it to their mouth. Others use legs to swim or fly.

KEY
- Tibia/Fibula
- Tarsi
- Metatarsi
- Phalanges

UNGULIGRADE I
HORSES
If you observe their footprints, you will see that only their hooves leave marks. Horses' hooves are made up of only one toe.

UNGULIGRADE II
GOATS
The majority of ungulates, such as goats, have an even number of toes. They are called artiodactyls as opposed to perissodactyls, which have an odd number of toes.

LYING FOOTPRINTS
Other species of unguligrades (or simply ungulates) can have more toes that make up their hooves, but they do not place weight on more than two of them.

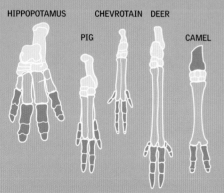

HIPPOPOTAMUS   CHEVROTAIN   DEER   PIG   CAMEL

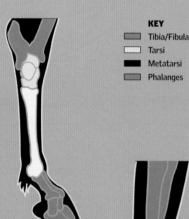

5 toes
THE NORMAL NUMBER FOR MAMMALS: RUNNING SPECIES HAVE FEWER.

DIGITIGRADE
DOG
These mammals place the full surface of their toes (or some of them) on the ground when walking. They usually leave the mark of their front toes and a small part of the forefoot as a footprint. Dogs and cats are the best-known examples.

PLANTIGRADE
HUMAN
Primates, including humans, bear their weight on their toes and much of the sole of the foot when walking, particularly on the metatarsus. Rats, weasels, bears, rabbits, skunks, raccoons, mice, and hedgehogs are also plantigrades.

SECOND TOE

NAIL

BIG TOE

LEFT FOOT OF CHIMPANZEE
Pan troglodytes
Life-size photo

DISTAL PHALANX

MEDIAL PHALANX

PHALANX

METATARSAL

SMALL

TARSI

WALK OR CLIMB
There is a fundamental difference between the human foot and that of a monkey. The monkey has a long, prehensile digit in its foot similar to that in its hand. Monkeys use their feet to grab branches as they move through the trees.

CHIMPANZEE   HUMAN

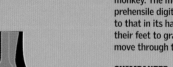
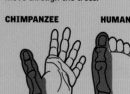
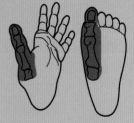

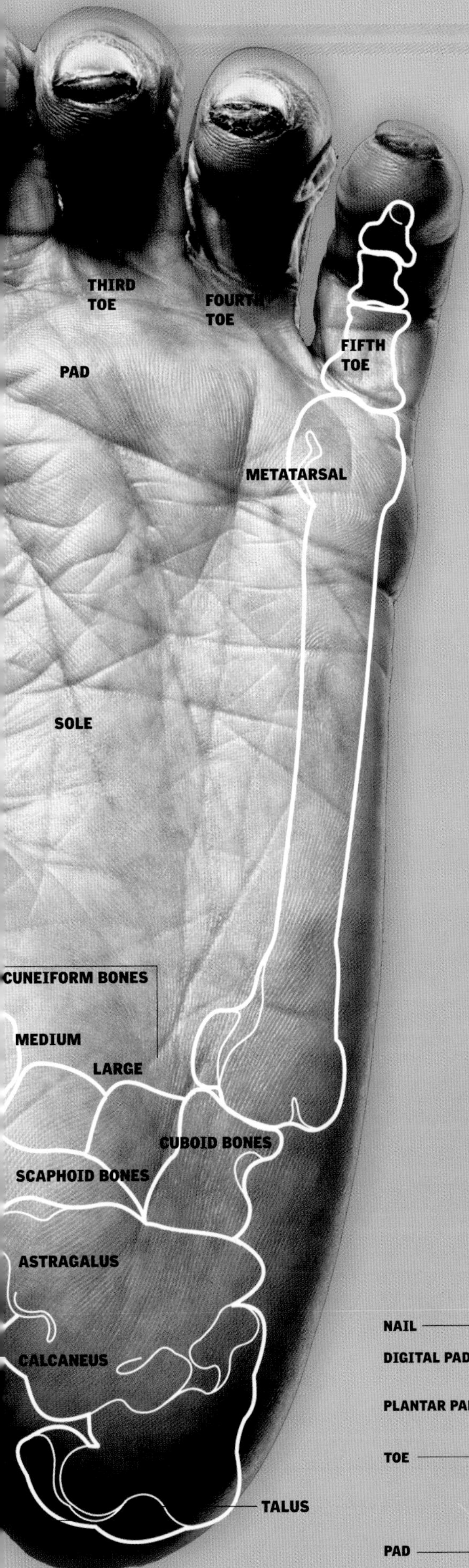

Chiroptera

From the Greek, meaning "winged hand," this is how bats are designated because their forelimbs are modified, the fingers thinning and lengthening to be able to support a membrane that functions as a wing. The hind limbs did not change similarly: they have claws.

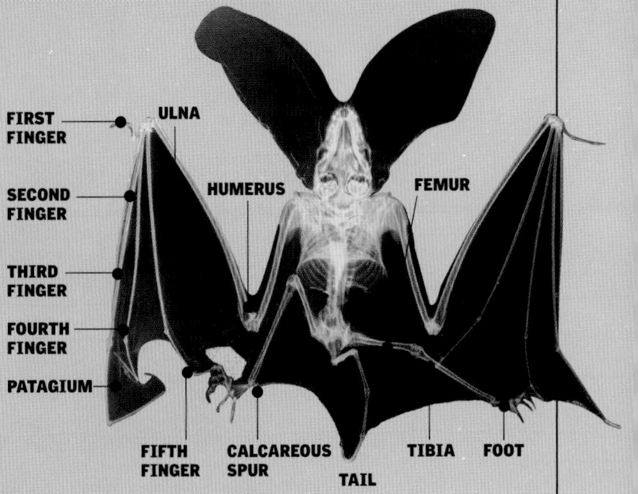

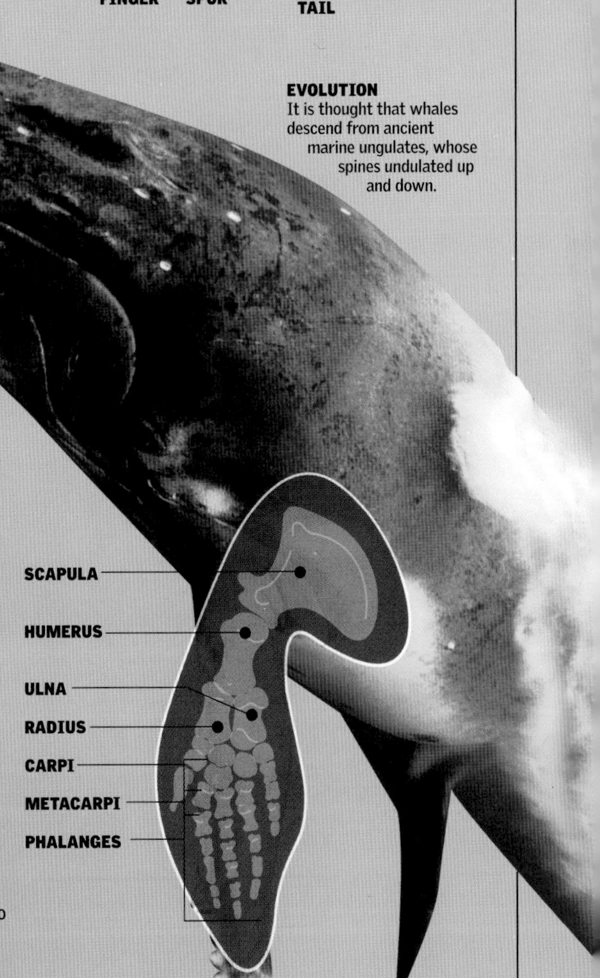

FIRST FINGER
ULNA
SECOND FINGER
HUMERUS
FEMUR
THIRD FINGER
FOURTH FINGER
PATAGIUM
FIFTH FINGER
CALCAREOUS SPUR
TIBIA
FOOT
TAIL

EVOLUTION
It is thought that whales descend from ancient marine ungulates, whose spines undulated up and down.

THIRD TOE
FOURTH TOE
PAD
FIFTH TOE
METATARSAL
SOLE

CUNEIFORM BONES
MEDIUM
LARGE
CUBOID BONES
SCAPHOID BONES
ASTRAGALUS
CALCANEUS
TALUS

Cetaceans

Whales adapted so well to the sea that they seem to be fish. But inside their fins—modified front legs—there is a bony structure similar to that of a hand with fingers. They have no hind limbs; the tail, placed horizontally and used to move in the water, has no connection to those limbs.

Tail

HORIZONTAL IN MAMMALS THAT SWIM, AS DISTINCT FROM FISH

Felines

The function of their paws is to support their agile and elastic bodies, allowing them to move about. The front paws also help in hunting to catch and hold prey.

SCAPULA
HUMERUS
ULNA
RADIUS
CARPI
METACARPI
PHALANGES

NAIL
DIGITAL PAD
PLANTAR PAD
TOE
PAD

RETRACTABLE NAIL

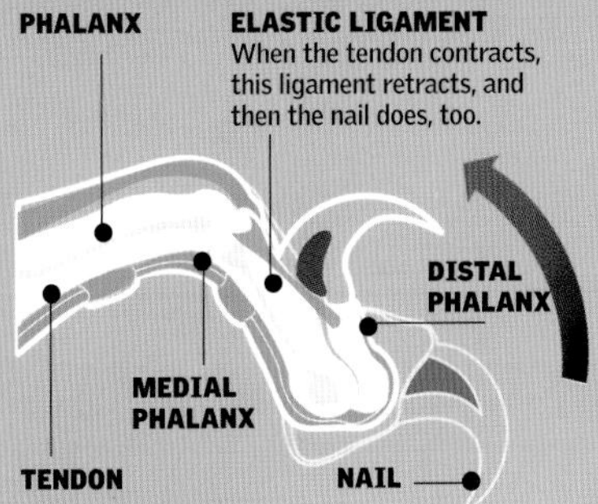

PHALANX
ELASTIC LIGAMENT
When the tendon contracts, this ligament retracts, and then the nail does, too.

DISTAL PHALANX
MEDIAL PHALANX
TENDON
NAIL

What Doesn't Run, Flies

They are meteors of flesh, bone, and hot blood. Cheetahs are the fastest of the land animals and unique members of the Felidae family, which hunt using their keen vision and great speed. They can reach more than 70 miles per hour (115 km/h) in short runs and reach 45 miles per hour (72 km/h) in an average of only 2 seconds. They can get above 60 miles per hour (100 km/h), but they can sustain that speed for only a few seconds. They look like leopards, although their physical characteristics are different: they are longer and thinner, and their heads are smaller and rounded.

TAKEOFF
From the top of a tree, it jumps toward another shorter tree.

Cheetahs

Whereas tigers prefer to lie in wait for prey and then jump on it, the cheetah uses explosive speed of more than 60 miles per hour (100 km/h) to run its prey down.

 Start
The cheetah begins running by lengthening its stride and extending its four legs.

 Spinal Contraction
Then it gathers its legs under its body, contracting its cervical spine to the maximum.

NOSTRILS
Very wide, they allow it to receive more oxygen as it runs.

ORDER	Carnivora
FAMILY	Felidae
SPECIES	*Acinonyx jubatus* (Africa)
	Acinonyx venaticus (Asia)

FIRST POINT OF CONTACT
As it runs, only one leg touches the ground at a time, but during the cervical contraction, the entire body lifts from the ground.

SECOND POINT OF CONTACT
Extending its four legs again, it picks up more momentum, supporting itself on only one back leg.

Bipeds Versus Quadrupeds

18 MPH (29 KM/H)
SIX-LINED RACERUNNER
Cnemidophorus sexlineatus

23 MPH (37 KM/H)
HUMAN BEING
Track record: Asafa Powell (Jamaica), 110 yards (100 m) in 9.77 seconds

42 MPH (67 KM/H)
GREYHOUND
A dog with a light skeleton and aerodynamic anatomy

50 MPH (80 KM/H)
HORSE
An anatomy designed for running, powerful musculature

70 MPH (115 KM/H)
CHEETAH
It only takes 2 seconds to reach a speed of 45 miles per hour (72 km/h).

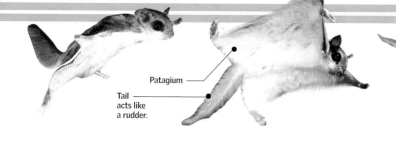

Patagium

Tail acts like a rudder.

TOES
Upon landing, it grabs onto the surface with its toes.

Siberian Flying Squirrel

Flying squirrels (*Pteromys volans*) belong to the same rodent family as common squirrels, to which they are similar in both appearance and way of life. They live in the mixed forests of northern Europe, across Siberia, and into East Asia.

IN THE AIR
The flying squirrel does not actually fly—it glides. Between its front and back limbs is a membrane of skin that, like a delta wing, stretches out the moment the animal jumps and stretches its legs. Thanks to that it can glide from the top of one tree to the trunk of another.

LANDING
While gliding, the squirrel can change its landing angle. Just before landing, it lowers its tail and raises its front legs, using the membrane like an air brake. It lands very gently on all four paws.

TAIL
Large compared to the rest of the body, it acts as a pivot used to suddenly change direction.

③ Extending the Spine
In a counterthrust opposing the contraction, the spine extends, creating forward momentum. The cheetah can cover 26 feet (8 m) in a single stride.

70 miles per hour
(115 km/h)
MAXIMUM SPEED, BUT CAN BE MAINTAINED FOR ONLY 550 YARDS (500 M)

SHOULDER
The extensive flexion of the shoulder allows it to take very long leaps.

HEAD
Small and aerodynamic, with low air resistance.

LIMBS
Long and agile. It has a powerful, flexible skeleton and musculature.

ZIGZAGGING AT HIGH SPEED

1 Cheetahs can make sharp turns while running at high speed.

2 These movements are possible because its nails are not retractable, so that cheetahs firmly grip the ground.

PAWS

DIGITS
5 in the hands
4 in the feet

NAILS
Unlike other felines, their nails are not retractable, allowing them to grip the ground better.

Sloth

These animals are notable for their extremely slow metabolism. They take half a minute to move a limb! They are also somewhat myopic, their hearing is mediocre, and their sense of smell barely serves to distinguish the plants on which they feed. They are at the extreme opposite of cheetahs. However, since they practically live perched in trees, they do not need to move or see or hear precisely. They are perfectly adapted to their way of life.

THREE-TOED SLOTH
Native to the Amazon River basin

Developed Senses

Dogs have inherited from wolves great hearing and an excellent sense of smell. Both perform an essential role in their relationship to their surroundings and many of their social activities. However, they are very dependent on the keenness of their senses, depending on the habitat in which they develop. Whereas humans often remember other people as images, dogs do so with their sense of smell, their most important sense. They have 44 times more olfactory cells than people do, and they can perceive smells in an area covering some 24 square inches (150 sq cm). Dogs can discern one molecule out of a million other ones, and they can hear sounds so low that they are imperceptible to people.

Hearing

The auditory ability of dogs is four times greater than that of human beings, and it is highly developed. Their ability depends on the shape and orientation of their ears, which allow them to locate and pay closer attention to sounds, although this varies by breed. They can hear sharper tones and much softer sounds, and they can directly locate the spatial reference point where a noise was produced. Dogs hear sounds of up to 40 kilohertz, whereas the upper limit for human hearing is 18 kilohertz.

INSIDE THE COCHLEA

Reissner's membrane
Scala vestibuli
Organ of Corti
Scala tympani

AURICULAR CARTILAGE

LABYRINTH
SEMICIRCULAR CANALS

AUDITORY NERVE

AUDITORY OSSICLES

INCUS (ANVIL)
MALLEUS (HAMMER)
STAPES (STIRRUP)

COCHLEAR NERVE

AUDITORY CANAL

MIDDLE EAR

COCHLEA

AUDITORY CANAL
TYMPANIC MEMBRANE

Dome

INTERNAL STRUCTURE OF THE BULLA

Crest

Ciliary Cells

The dome diverts sounds toward the bulla and other organs that direct electric signals to the brain.

OVAL WINDOW

EUSTACHIAN TUBE

COCHLEA

AUDITORY LEVELS

	0 hertz	1	10	100	1,000	10,000	20,000	40,000
PEOPLE								
FOXES								
MICE								
BATS								
FROGS								
ELEPHANTS								
BIRDS								

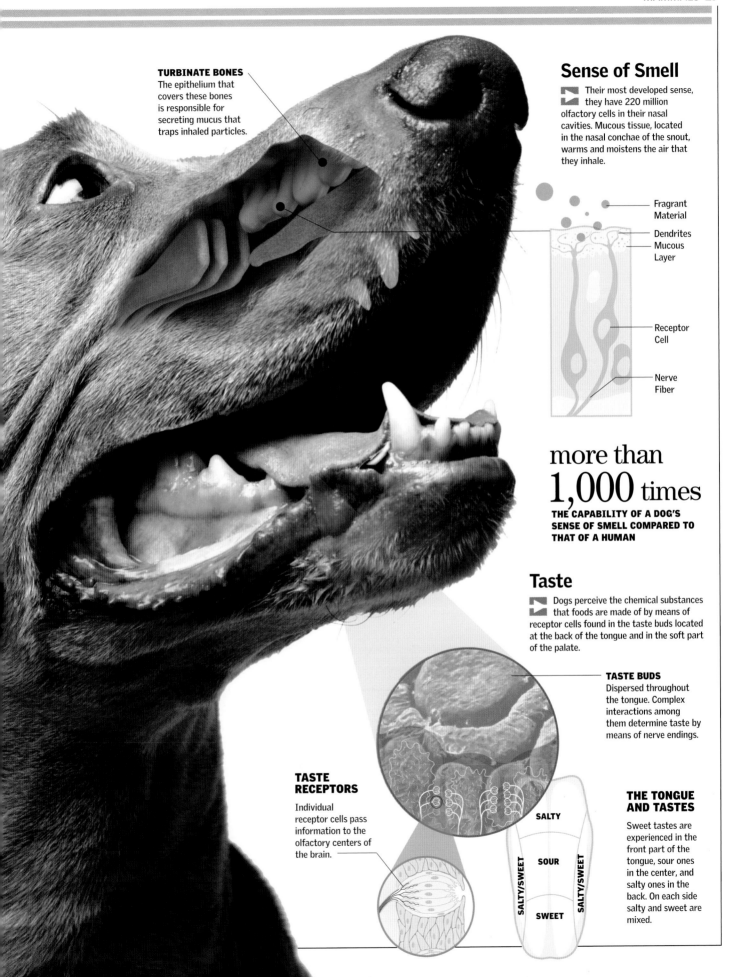

TURBINATE BONES
The epithelium that covers these bones is responsible for secreting mucus that traps inhaled particles.

Sense of Smell

Their most developed sense, they have 220 million olfactory cells in their nasal cavities. Mucous tissue, located in the nasal conchae of the snout, warms and moistens the air that they inhale.

Fragrant Material

Dendrites

Mucous Layer

Receptor Cell

Nerve Fiber

more than 1,000 times

THE CAPABILITY OF A DOG'S SENSE OF SMELL COMPARED TO THAT OF A HUMAN

Taste

Dogs perceive the chemical substances that foods are made of by means of receptor cells found in the taste buds located at the back of the tongue and in the soft part of the palate.

TASTE BUDS
Dispersed throughout the tongue. Complex interactions among them determine taste by means of nerve endings.

TASTE RECEPTORS

Individual receptor cells pass information to the olfactory centers of the brain.

SALTY

SALTY/SWEET

SOUR

SALTY/SWEET

SWEET

THE TONGUE AND TASTES

Sweet tastes are experienced in the front part of the tongue, sour ones in the center, and salty ones in the back. On each side salty and sweet are mixed.

Soft Contact

Admired, adored, and coveted by humans, a mammal's fur coat is much more than a skin covering. It acts as a protective layer against mechanical injuries, prevents invasion by germs, and regulates the loss of body heat and moisture. In many species, such as the Arctic fox, it provides camouflage by changing color and texture from winter to summer.

Fur and Mimicry

Mammals from cold regions, such as polar bears, have white fur to camouflage themselves in snow. Others, such as polar, or Arctic, foxes and the American hare, change their fur color with the seasons, because they live in areas that are snow-covered in winter, where their brown summer fur would make them easy prey. Lions' beige color helps them avoid being discovered while they stalk their prey.

WINTER
Arctic foxes have two kinds of color phases. White phase foxes are almost pure white in the winter, which allows them to camouflage themselves in the snow and ice.

SUMMER
The fur coat of the Arctic fox (*Alopex lagopus*) in summer is half as thick as that of winter, with less than half the underfur. In summer, "white" phase animals turn a gray-brown to grayish color, and those that have a "blue" phase are browner and darker.

UV **FUR SERVES TO PROTECT THE SKIN FROM EXCESSIVE UV RAYS.**

The Skin

EPIDERMIS
Outer layer formed by resistant, flat cells

DERMIS
Layer with blood vessels, glands, and nerve endings. It is a layer of sebaceous glands that secrete an oily substance, sebum, on the surface of the skin.

RUFFINI'S CORPUSCLE

FATTY TISSUE
This is a specialized conjunctive tissue made up primarily of connective cells called adipocytes, which store energy in the form of triglycerides.

SWEAT GLANDS
When the body is hot, the glands secrete sweat, which passes through the sweat ducts to the surface of the skin.

HAIR SHAFT

SWEAT PORE

STRATUM CORNEUM

ARRECTOR PILI MUSCLE

FOLLICLE

ARTERY

VEIN

GRAY WOLF

HARE

CHINCHILLA   MACAQUE MONKEY

HAIR STRUCTURE

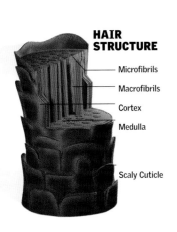

Microfibrils
Macrofibrils
Cortex
Medulla
Scaly Cuticle

Diverse Hairs

The majority of mammals' fur is made up of more than one type of hair, and its different colors are due to a group of proteins called melanins. Each coat has different layers. Guard hairs are the first layer, providing protection. Underneath that, there is a fine layer called underfur, formed by constantly growing short hairs that renew the coat.

BAT HAIR
Each strand of hair has an outer cuticle formed by superposed scales.

POLAR BEAR HAIR
Each one of its hairs is hollow and filled with air. This heightens the insulating capability of the inner layer.

WOOL FIBER

Protofibril
Microfibril
Macrofibril
Cortex 90%
Cuticle 10%

ENLARGED WOOL
This is the most complex natural textile fiber in existence. It absorbs moisture but repels water.

PORCUPINE QUILLS

Called guard hairs, they are located outside the fur. In the case of the porcupine, they have been modified to form defensive quills.

30,000

THE NUMBER OF QUILLS THAT COVER A PORCUPINE (148 PER SQUARE INCH/23 PER SQ CM)

MINI-QUILLS
Sharp scales

DERMAL PAPILLA
attaches the dermis to the epidermis.

MERKEL'S DISK
A sense receptor under the skin's surface that responds to light, continuous touch and pressure

SEBACEOUS GLAND
secretes a waxy substance, or sebum, which moistens the skin, making it waterproof.

PACINIAN CORPUSCLE
Sense receptors under the dermis. The Pacini receptors lie under the layer of deep fat and detect vibration and pressure.

Insulating Skin

Insulation is one of the functions of animals' skins and hair. It not only helps to conserve body warmth but also, as in the case of camels, protects them from excessive heat. Its color often blends in with its surroundings, serving as camouflage.

OUTER FUR

UNDERFUR

LAYER OF FAT

ERECTION MECHANISM

Base of the Quill
Epidermis

Connective Tissue

Root

Retinaculum

1 When the quill touches a strange surface, it exerts a light downward pressure on the epidermis.

2 The fine tissue that covers the root of the quill breaks.

3 The erector pili muscle receives the contact signal and contracts.

COATI

SEA LION (JUVENILE)

PORCUPINE

Behavior and Life Cycle

M ammalian reproduction is sexual and by internal fertilization, which involves copulation between the male and the female. Mammals are also characterized by the offspring's dependence on its parents. In any case, there is a group of mammals called monotremes that is oviparous; that is, its members reproduce

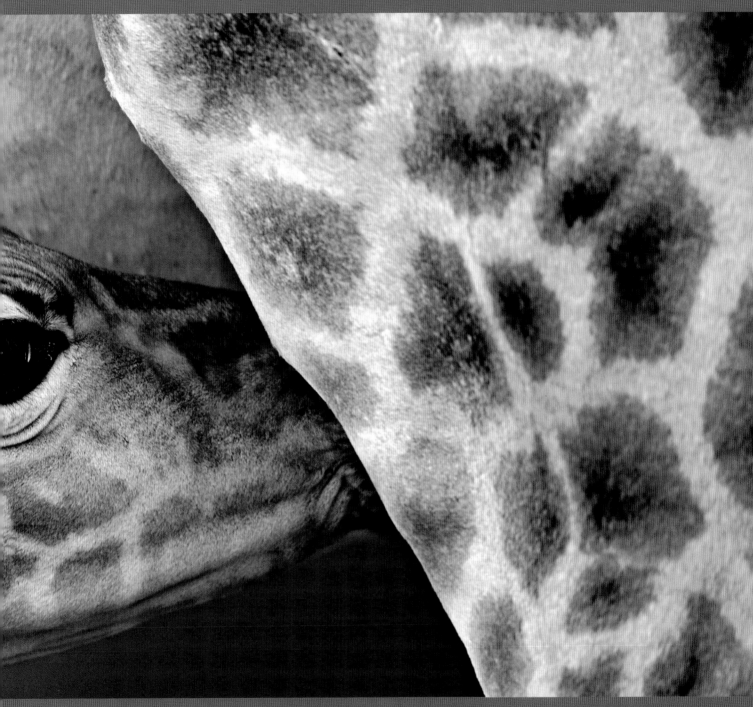

by laying eggs. Mammalian behavior consists of a mixture of inherited components and components that can be shaped by learning. Part of this process is accomplished through play, since the young use such encounters to practice jumping, biting, hunting, and other survival skills. You will discover this and much more when you turn the page.

Life Cycle

Birth, maturity, reproduction, and death: this life cycle has certain particularities among mammals. As a general rule, the larger a mammal, the longer the members of its species tend to live but the fewer offspring are born to a single female per litter or reproductive season. Most mammals, including humans, are placental mammals; their vital functions are fully developed inside the body of the mother.

90 Years

A WHALE'S AVERAGE LIFE SPAN— THE GREATEST OF ANY LIVING MAMMAL

Placental Mammals

This is the largest group of mammals, the one that has multiplied most on the planet, although its form of gestation and lactation produces great wear and tear on the females, making them less prolific. They are generally polygamous: a few males (the most competitive) fertilize many females, and other males, none. Only 3 percent of mammals are monogamous in each season. In these cases, males participate in rearing the offspring, as they also do when resources are scarce. If resources are abundant, the females take care of the young alone, and the males mate with other females.

They make use of natural caves or dig underground.

Weaning
35 TO 40 DAYS

Young rabbits remain with their mother even after nursing ends for protection and the inculcation of species-specific behavior.

Sexual Maturity
5 TO 7 MONTHS

The better rabbits are fed, the more quickly they become capable of reproducing. They are considered adults at 8 or 9 months, when they weigh some 2 pounds (900 g).

They have four to five pairs of breasts.

Lactation
25 TO 30 DAYS

fed upon milk, although they can digest solid food after 20 days. The young abandon the burrow after 35 or 40 days and remain in the area where they were raised (philopatry).

Female rabbits can mate at any time.

Gestation
28 TO 33 DAYS

They spend it in a collective burrow (warren) dug in the ground and covered with vegetation and fur. The female will abandon it as soon as lactation ends.

EASTERN COTTONTAIL RABBIT
Sylvilagus floridanus

Longevity
4 to 10 years

4 inches (10 cm)

They are born without fur, with semitranslucent skin.

AT BIRTH
The young weigh some 1½ to 1¾ ounces (40–50 g). They do not open their eyes until the tenth day.

NUMBER OF OFFSPRING

In general, it is inversely proportional to the species' size.

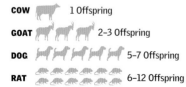

COW	1 Offspring
GOAT	2–3 Offspring
DOG	5–7 Offspring
RAT	6–12 Offspring

3 to 9 Young

PER LITTER, AND FROM 5 TO 7 LITTERS PER YEAR

Marsupials

Very short gestation period, after which they develop in a kind of partially open pouch (the marsupium), which the female carries on her belly. The majority of the roughly 300 known species of marsupials are solitary, except in mating periods. In general, they are promiscuous animals, although some, such as wallabies (small kangaroos), tend to mate with the same female all their life.

Lactation
22 WEEKS

A muscle inside the pouch prevents the infant from falling out. At 22 weeks, it opens its eyes, and a type of pap produced by its mother is added to its diet, which will prepare it for a herbivorous diet.

Gestation
35 DAYS

With its extremities and functional organs barely developed at birth, the newborn must crawl by itself from the cloaca to the pouch to continue its development.

The young animal fastens itself to its mother and is carried around by her, clinging to her shoulders.

BANISHED OFFSPRING
Dominant males keep the offspring and other young males apart.

Dominant males mate with all the females.

Some females leave to look for strong males.

KOALA
Phascolarctos cinereus

By the end of lactation, fur covers the whole body.

¾ inch (2 cm)

Leaving the Pouch
1 YEAR

The offspring reaches a size that allows it to fend for itself. It has already incorporated herbivorous food into its diet. The mother can become pregnant again, but its young will remain nearby.

Sexual Maturity
3 TO 4 YEARS

At two years, koalas already have developed sexual organs (females earlier than males). But they do not start mating until one or two years later.

LONGEVITY

PEOPLE	70 years
ELEPHANTS	70
HORSES	40
GIRAFFES	20
CATS	15
DOGS	15
HAMSTERS	3

1 offspring
1 BIRTH PER YEAR

Longevity
15 to 20 years

GESTATION PERIODS

ANIMAL	MONTHS
Elephants	23
Giraffes	17
Gibbons	9
Lions	7
Dogs	2

COMPARISON OF EGG SIZE

The shell is soft and facilitates the offspring's birth. Unlike birds, they do not have beaks.

CHICKEN

ECHIDNA

Monotremes

Mammals whose females lay eggs are generally solitary species for most of the year. Platypuses are seen as couples only when they mate. Although they have a period of courtship for one to three months, the males have no relationship with the females after copulation or with the offspring. Short-beaked echidna females practice polyandry, copulating with various males in various seasons.

Incubation
12 DAYS

Eggs gestate for a month before hatching. They incubate within a pouch for about ten days to remain at the proper temperature until the young are born.

Newborn Offspring

Undeveloped Limbs

Shell

½ inch (15 mm)

1 to 3
EGGS AT A TIME

In the Pouch
2 TO 3 MONTHS

After breaking the shell, the young are suckled while they remain in a kind of pouch of the female.

Underground cave or a cave among rocks

The fur is already spiny.

Weaning
4 TO 6 MONTHS

After three months, the offspring can leave the burrow or remain in it alone for up to a day and a half before finally separating from the mother.

Longevity
50 years

SHORT-BEAKED ECHIDNA
Tachyglossus aculeatus

Beauty and Height

Finding a female with whom to mate is the great effort of the male's life, a competition with other males of his own species. Each animal has its particular nuances. For stags, antlers play a fundamental role in winning the heart of their chosen one. Whichever stag has the most beautiful, longest, and sharpest horns will be the winner. Thus, he will be able to defend his territory, court the female, and reproduce.

ANTLER LAYERS

EPIDERMIS

PERIOSTEUM Fibrous tissue that protects the bone

DERMIS

Red Deer

These are svelte, robust, well-formed animals with a majestic and haughty carriage. They are very timid and fearful, and it is thought that the species is 400,000 years old. They are active at daybreak and evening, and males usually live alone. Females and younger deer group in herds.

ORDER	Artiodactyla
FAMILY	Cervidae
SPECIES	*Cervus elaphus*
DIET	Herbivorous
WEIGHT (MALE)	400 pounds (180 kg)

Fights

When two males fight over a harem, each will display his antlers to frighten his rival. The horns can also be used to defend against predators.

24 inches (60 cm)

MALE
43 inches (110 cm)

FEMALE
31 inches (80 cm)

4 NEW
Near the end of summer, stags display their new antlers, which will be larger and heavier than the previous ones.

Molt
Antlers are shed every year. Animals between 6 and 10 years old display the finest antlers.

3 DEVELOPMENT
Stags rub their antlers against trees and bushes to get rid of the membrane that covers them.

1 FALLING OFF
At the onset of fall, stags begin to lose their antlers, which will be replaced by new ones.

2 GROWTH
New antlers are covered with a fine membrane called velvet. Velvet will remain on the horns until they are fully developed.

Antlers

FORK   PALM   POINT

BEAM

CROWN

PEDICLE

Horns and Antlers
Horns are outgrowths of the cranium, covered by a tegument that forms a sheath. They appear in bovids of both sexes and are generally permanent. Antlers are also extensions of the cranium; they are limited to the deer family, are present only in males, and are replaced annually.

Bellows
Sonorous and discordant, they begin to be heard when spring arrives, announcing the beginning of rut, or mating season. They not only attempt to keep competitors away with their call but they also use the sound to attract unattached females to join the male's herd.

Oviparous Mammals

For a mammal to lay eggs seems improbable, but the surprising monotreme females, instead of giving birth to young, are oviparous. They are warm-blooded, have hair, and feed their newborn through mammary glands despite having no nipples. Platypuses seem like a cocktail of nature, inasmuch as parts of their bodies resemble those of other types of animal. The other monotremes, echidnas, are covered with spines, and their young grow in the mother's pouch.

Platypus

Combining the skin of a mole, the tail of a beaver, the feet of a frog, and the beak of a duck, platypuses are semiaquatic mammals endemic to the eastern part of Australia and to the island of Tasmania. They construct burrows in riverbanks consisting of a long passageway.

FAMILY	Ornithorhynchidae
SPECIES	*Ornithorhynchus anatinus*
DIET	Herbivorous
WEIGHT	5½ pounds (2.5 kg)

16 To 24 inches
(40–60 CM)

BILL
has sensitive electroreceptors that can perceive the electric field generated by the muscles of their prey.

100 feet
(30 m)

HOW LONG THE BURROW OF A PLATYPUS CAN BE

Echidna

Lives in Australia, New Guinea, and Tasmania. It has an elongated snout in the form of a beak, no teeth, and a long, retractable tongue. It is a notable digger and hibernates underground. Echidnas can live up to 50 years, and their hair varies according to the species.

FAMILY	Tachyglossidae
SPECIES	*Tachyglossus aculeatus*
ADULT SIZE	

12 To 35 inches
(30 to 90 cm)

RETRACTABLE TONGUE
A sticky substance on the long and slender tongue allows it to catch termites and ants.

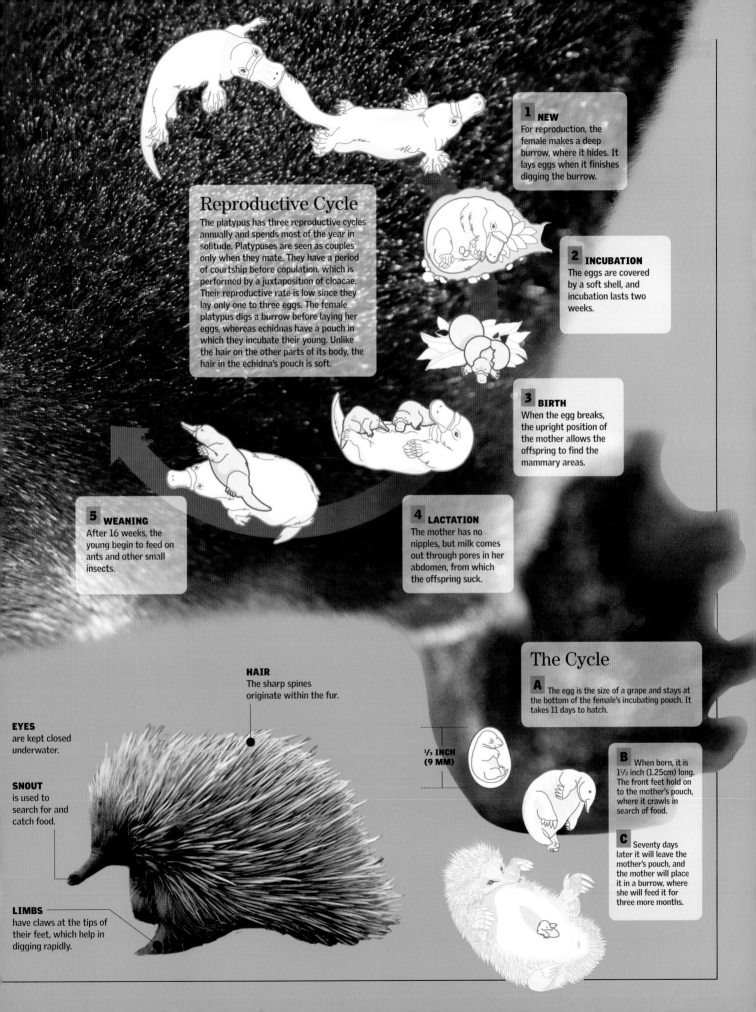

1 NEW
For reproduction, the female makes a deep burrow, where it hides. It lays eggs when it finishes digging the burrow.

Reproductive Cycle

The platypus has three reproductive cycles annually and spends most of the year in solitude. Platypuses are seen as couples only when they mate. They have a period of courtship before copulation, which is performed by a juxtaposition of cloacae. Their reproductive rate is low since they lay only one to three eggs. The female platypus digs a burrow before laying her eggs, whereas echidnas have a pouch in which they incubate their young. Unlike the hair on the other parts of its body, the hair in the echidna's pouch is soft.

2 INCUBATION
The eggs are covered by a soft shell, and incubation lasts two weeks.

3 BIRTH
When the egg breaks, the upright position of the mother allows the offspring to find the mammary areas.

5 WEANING
After 16 weeks, the young begin to feed on ants and other small insects.

4 LACTATION
The mother has no nipples, but milk comes out through pores in her abdomen, from which the offspring suck.

The Cycle

A The egg is the size of a grape and stays at the bottom of the female's incubating pouch. It takes 11 days to hatch.

HAIR
The sharp spines originate within the fur.

EYES
are kept closed underwater.

SNOUT
is used to search for and catch food.

LIMBS
have claws at the tips of their feet, which help in digging rapidly.

⅓ INCH (9 MM)

B When born, it is 1½ inch (1.25cm) long. The front feet hold on to the mother's pouch, where it crawls in search of food.

C Seventy days later it will leave the mother's pouch, and the mother will place it in a burrow, where she will feed it for three more months.

Efficient Nursery

Marsupial females carry their newborn offspring in their marsupium, a pouch attached to their belly. The offspring are not very well developed when they come into the world after a gestation period that varies from two to five weeks. Upon emerging, the offspring must immediately climb with their front paws to the marsupium to survive. Once inside, they will be protected. They are continually supplied with milk through their mother's four teats, helping them complete their growth before leaving the pouch for the outside world.

Red Kangaroos

Kangaroos are a family comprising several groups, including great wallabies and tree-dwelling kangaroos. Kangaroos, the prototypical marsupial, live in Australia and in Papua New Guinea, never more than 9 miles (15 km) from water. They have large, muscular hind legs that they use to take great consecutive leaps, reaching speeds of 15 to 20 miles per hour (24–32 km/h). They are able to maintain their balance standing only on their hind legs. Their heel bone (calcaneus) is long and acts as a lever.

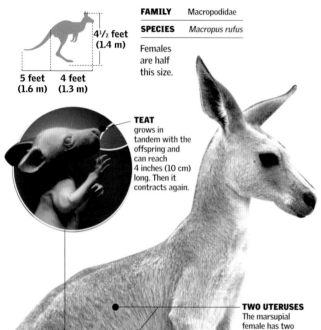

4½ feet (1.4 m)
5 feet (1.6 m)
4 feet (1.3 m)

FAMILY Macropodidae
SPECIES *Macropus rufus*

Females are half this size.

TEAT grows in tandem with the offspring and can reach 4 inches (10 cm) long. Then it contracts again.

TWO UTERUSES The marsupial female has two uteruses.

REPRODUCTIVE CYCLE

0 DAYS BIRTH OF THE KANGAROO
237 DAYS A NEW KANGAROO IS BORN
2 DAYS RUT AND NEW CONCEPTION
236 DAYS THE OFFSPRING BECOMES INDEPENDENT
238 DAYS RUT AND NEW CONCEPTION

The female can give birth to an offspring while another one is in the marsupium.

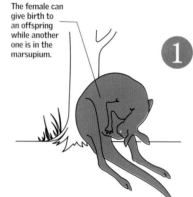

1 Smoothing the Way

When preparing for the birth of an offspring, the female kangaroo licks its coat to form a kind of path some 5½ inches (14 cm) long, which the offspring will follow to reach the entrance to the pouch located higher up on the belly.

2 A Marathon

Small kangaroos are born after a few weeks of gestation in an early stage of their development, weighing less than ¼ ounce (5 g). They cannot see or hear. They only move their front paws, with which they drag themselves, following their mother's trail of saliva and guided by their sense of smell.

The baby kangaroo must get to the pouch within three minutes or it will not survive.

MOVING OUT OF THE MARSUPIUM
At eight months, the offspring leaves the pouch and begins to add grass to its diet, but it will continue to be suckled until it is 18 months old.

3 Lactation

Upon reaching the marsupium, the baby fastens its mouth upon one of the four teats inside. At this point, the baby is red and looks very fragile. However, it will grow continuously over the next four months, during which it will not leave the pouch.

Entering the Marsupium

A After some eight months, the kangaroo can leave the marsupium. But it returns to be suckled and protected.

B However, because it barely fits, it enters head first with the aid of its front paws and turns around once inside the pouch.

C When it is already alternating milk with grass from outside, the young kangaroo sticks its head out to eat grass without leaving the pouch.

¾ inch
(20 mm)

**THE SIZE OF AN OFFSPRING WHEN
IT ENTERS THE MARSUPIUM**

Miraculous Placenta

The largest reproductive group is formed by placental mammals, in which the unborn offspring develop in the female's uterus. During gestation, food and oxygen pass from the mother to the fetus through an organ known as the placenta, which allows the exchange of substances through the blood. At birth, the offspring often have no hair, are deaf and blind, and feed on milk secreted by the female's mammary glands, which become active after birth.

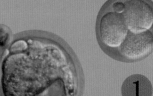

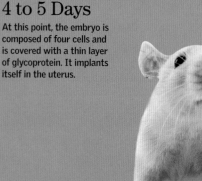

1
1 to 2 Days
Rat embryo at the two-cell stage. By the second day, it will have four cells, and on the third day, it will enter the uterus.

2
4 to 5 Days
At this point, the embryo is composed of four cells and is covered with a thin layer of glycoprotein. It implants itself in the uterus.

Gestation of Rats

Gestation lasts between 22 and 24 days. Whereas the placenta is discoid and hemochorial, the ovaries are essential for maintaining gestation. If an ovariectomy is performed at any stage of gestation, it will always bring about a miscarriage or the reabsorption of the fetuses because the placenta does not produce sufficient progesterone to maintain gestation. The growth of the uterine horns becomes visible on the thirteenth day of gestation.

YOLK
Implanted blastocyst, with trophoblastic cone and inner cell mass

3
6 to 8 Days
The blastocyst has now implanted and established itself in the uterus. The fetus begins to form, and the blastocyst becomes a yolk sac.

EYE
begins to develop and can now be observed.

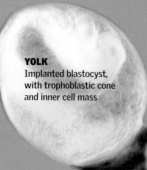

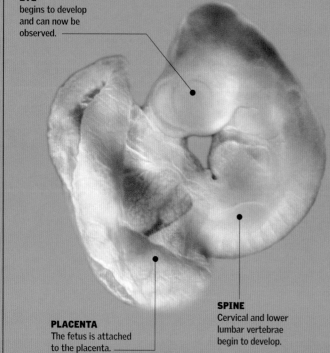

4
11½ Days
The embryo has now fastened itself to the embryonic sac (a kind of balloon that covers the fetus) and to the placenta. The brain, eyes, and legs begin to form.

BRAIN
The brain is forming; it appears transparent.

ORGANS
Internal organs begin to form and become visible.

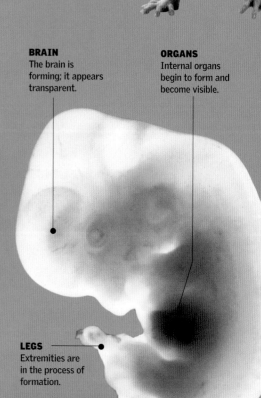

5
14½ Days
Eyes and extremities are now visible, and the internal organs begin to develop. A pre-cartilaginous maxillary and the outer ear begin to form.

LEGS
Extremities are in the process of formation.

PLACENTA
The fetus is attached to the placenta.

SPINE
Cervical and lower lumbar vertebrae begin to develop.

Placenta

From whales to shrews, placental mammals are characterized by gestating their young inside the mother and giving birth when they are well developed. To do so, they have a special organ, the placenta. This is a spongy tissue that completely surrounds the embryo, allowing for the exchange of substances through the blood. In this way, the mother can transfer nutrients and oxygen to the embryo, at the same time that she absorbs the metabolic waste of her future offspring. After birth, the placenta is immediately devoured by the mother, who uses her teeth to help the young leave the structure.

SPINE
The spine can be distinguished and is ready to support the little rat.

ORGANS
The organs are now almost complete and ready to go out into the world.

EYELIDS
They grow very rapidly, and by day 18, the eyes are already covered.

Uterus
IS BICORNUATE AND HAS TWO CERVICES.

TOES
Toes on the front limbs can also be distinguished.

7
19½ Days
Only a few days are left before the female will give birth to a new litter of little rats. At birth, they are helpless despite the fact that all their organs are developed.

³/₈ inch (10 mm)

⁵/₈ to ³/₄ inch (16 to 20 mm)

6
17½ Days
The eyelids grow very rapidly, and within a few hours the eyes will be completely covered. The palate has already completed its development, and the umbilical cord retracts.

The First Days

Mammals whose offspring develop within the uterus devote a lot of attention to their young compared to other animals, because their pups are unable to live on their own at birth. That is why they are cleaned, fed, and warmed. Dogs have various developmental stages. First is the neonatal stage, which lasts from the opening of the pups' eyes until they begin to hear. Then comes the socialization stage, which runs from days 21 to 70, and, finally, the juvenile stage, from 70 days on.

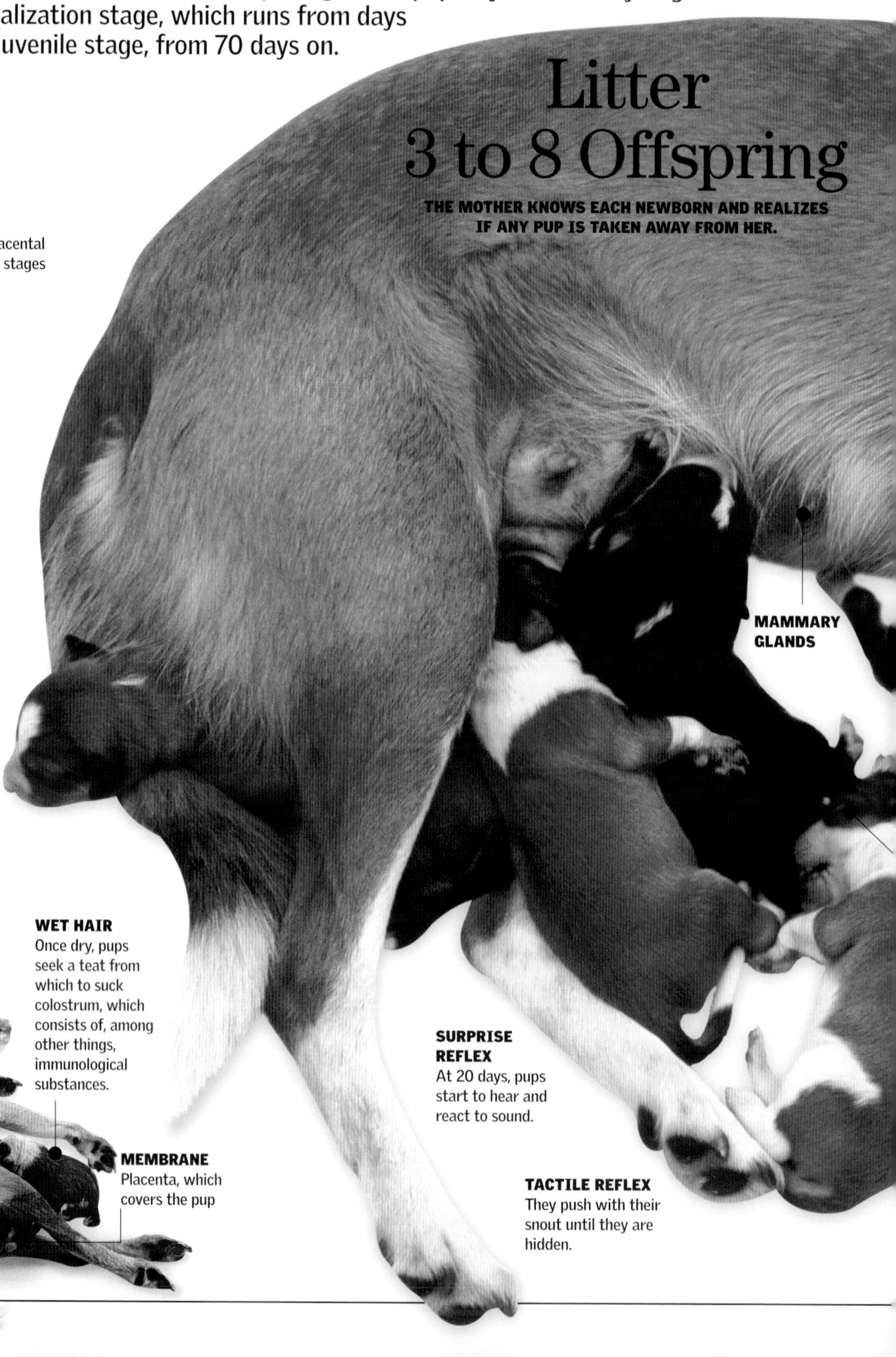

Litter
3 to 8 Offspring
THE MOTHER KNOWS EACH NEWBORN AND REALIZES IF ANY PUP IS TAKEN AWAY FROM HER.

Lactation Period

This period is essential in the reproductive process of mammals. The young of most placental mammals are totally dependent in the first stages of their life on mammary milk secretion.

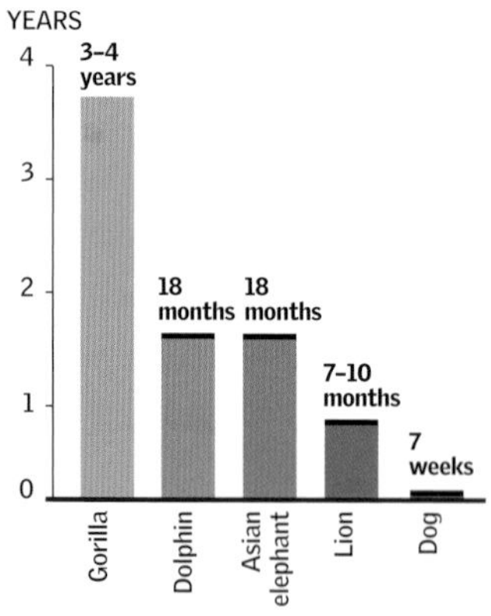

YEARS

Gorilla	3–4 years
Dolphin	18 months
Asian elephant	18 months
Lion	7–10 months
Dog	7 weeks

Birth

Like humans, dogs develop slowly after birth, because they are not fully developed when they come into this world and are incapable of living on their own. They need a structured environment in which they are cared for by their parents and other members of the pack.

MAMMARY GLANDS

BIRTH
The first pup is born between 1 and 2 hours after contractions begin.

WET HAIR
Once dry, pups seek a teat from which to suck colostrum, which consists of, among other things, immunological substances.

MEMBRANE
Placenta, which covers the pup

SURPRISE REFLEX
At 20 days, pups start to hear and react to sound.

TACTILE REFLEX
They push with their snout until they are hidden.

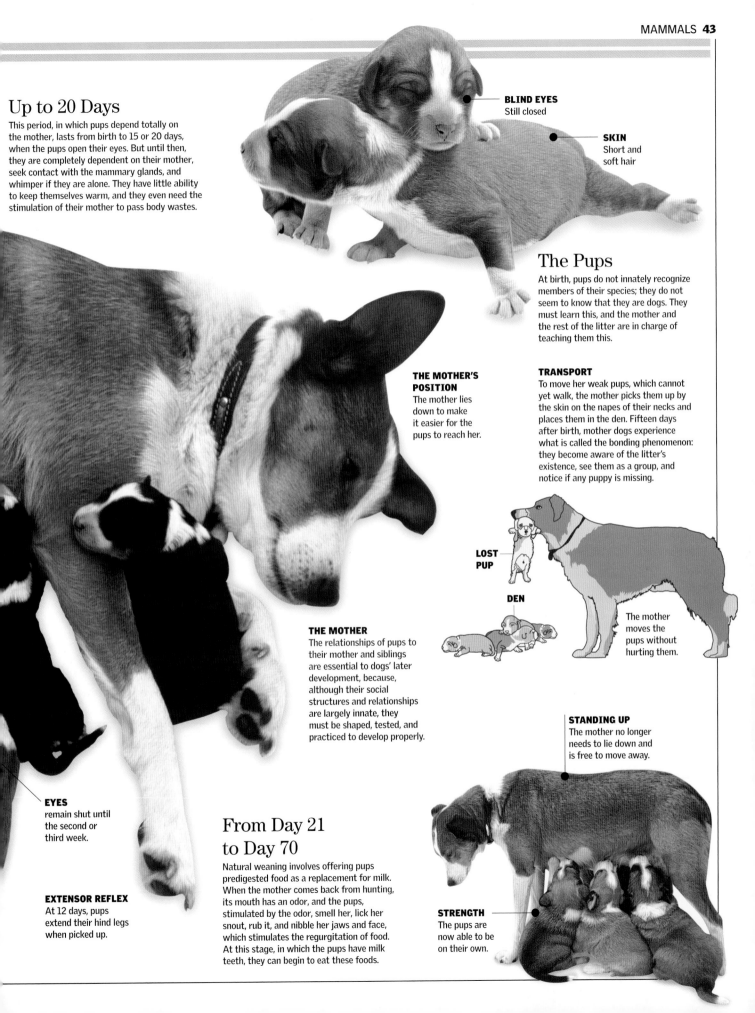

Up to 20 Days

This period, in which pups depend totally on the mother, lasts from birth to 15 or 20 days, when the pups open their eyes. But until then, they are completely dependent on their mother, seek contact with the mammary glands, and whimper if they are alone. They have little ability to keep themselves warm, and they even need the stimulation of their mother to pass body wastes.

BLIND EYES
Still closed

SKIN
Short and soft hair

The Pups

At birth, pups do not innately recognize members of their species; they do not seem to know that they are dogs. They must learn this, and the mother and the rest of the litter are in charge of teaching them this.

THE MOTHER'S POSITION
The mother lies down to make it easier for the pups to reach her.

TRANSPORT
To move her weak pups, which cannot yet walk, the mother picks them up by the skin on the napes of their necks and places them in the den. Fifteen days after birth, mother dogs experience what is called the bonding phenomenon: they become aware of the litter's existence, see them as a group, and notice if any puppy is missing.

LOST PUP

DEN

The mother moves the pups without hurting them.

THE MOTHER
The relationships of pups to their mother and siblings are essential to dogs' later development, because, although their social structures and relationships are largely innate, they must be shaped, tested, and practiced to develop properly.

EYES
remain shut until the second or third week.

STANDING UP
The mother no longer needs to lie down and is free to move away.

EXTENSOR REFLEX
At 12 days, pups extend their hind legs when picked up.

From Day 21 to Day 70

Natural weaning involves offering pups predigested food as a replacement for milk. When the mother comes back from hunting, its mouth has an odor, and the pups, stimulated by the odor, smell her, lick her snout, rub it, and nibble her jaws and face, which stimulates the regurgitation of food. At this stage, in which the pups have milk teeth, they can begin to eat these foods.

STRENGTH
The pups are now able to be on their own.

Development and Growth

Play is much more than entertainment for young mammals. This activity, which may appear to have no specific purpose, is the way in which they learn to be part of their species in the early stages of their lives, simultaneously acquiring the basic means of survival. In their games, chimpanzees perform primary instinctive activities that, with time and improvement, will become perfected instinctive activities. These include using tools, balancing in trees, and forming communication. Young chimpanzees express themselves by means of sounds, facial gestures, and body postures they imitate from adults. Play also allows them to develop their muscle strength and achieve good motor coordination.

More than 15

TYPES OF CALLS

are emitted by chimpanzees, including its pant-hoot: screams and grunts that can be heard a 1¼ mile (2 km) away. Pant-hoots are unique to the individual and can help to identify each member of the group.

This expression communicates terror.

This expression transmits submission.

This gesture indicates worry.

Communication

Some mammals, especially chimpanzees, communicate through facial expressions. This ability is well developed in the young primates, which express fear, submission, and worry, among other feelings.

Games

What we humans call play appears to be limited only to mammals, because they have well-developed senses, intelligence, and the ability to learn. It is through play that mammals carry out their learning.

Social Relations

Play also helps encourage apes to identify with their species. It provides a basis for learning to communicate through the use of sounds and body posture to express, for example, submission or domination.

IDENTIFICATION

Only 15 minutes of play with peers per day will moderate the effects of social isolation.

Survival

Play also functions as a method of learning to survive in a wild habitat. It trains carnivores in hunting techniques and herbivores in detecting, and fleeing from, danger.

Extremities

Chimpanzees are characterized by their long arms, which are endowed with great strength, and by their opposable thumbs. The digits of their hands and feet are large, allowing them to climb with great ease. They can hold onto a branch with their foot while they pluck its fruit with their hand.

Opposable thumb

Long digits

When they move around on all fours, they bear their weight on the soles of the feet and the knuckles of their hands.

Use of Tools

The use of tools is not common in mammals. However, chimpanzees are capable of using objects as tools, a skill they acquire by observing adults. They can use sticks to eat termites or use leaves as spoons to drink water.

Words

THEY CAN LEARN AND EXPRESS WORDS USING SIGN LANGUAGE.

PERCEPTION
They have sensory abilities very similar to those of people, and they distinguish smells better. Because of their large brains, they are very intelligent and can communicate with people by signs.

A chimpanzee pokes a stump in search of termites, using a stick as a tool.

A LIFE OF HANGING
A great entertainment for apes is hanging from trees. This exercise improves their coordination and arm strength.

Meat Eaters

The carnivore group is composed of species whose diet is based on hunting other animals. The kind of teeth they have help them efficiently cut and tear the flesh of their captured prey. Lions, the most sociable of the felines, have good vision and sharp hearing; they live in packs, and when they go hunting, they do so as a group.

Lions

are characterized by a strong, muscular physique. A male requires 15$\frac{1}{2}$ pounds of meat (7 kg) a day, whereas a female needs 11 pounds (5 kg). They have a short digestive tract, which rapidly absorbs nutrients from the ingested meat.

Teeth

UPPER PREMOLARS

UPPER INCISORS

UPPER CANINE

CARNASSIAL PAIR

They are very large, and the dental crowns are two long blades arranged as shears that fit into each other. Together they slice and cut flesh to perfection.

ANTERIOR PREMOLARS

LOWER CANINE

LOWER INCISORS

The Hunt

1 **LYING IN AMBUSH**
Hidden in the grass, the lioness silently approaches the prey. Other females wait in hiding.

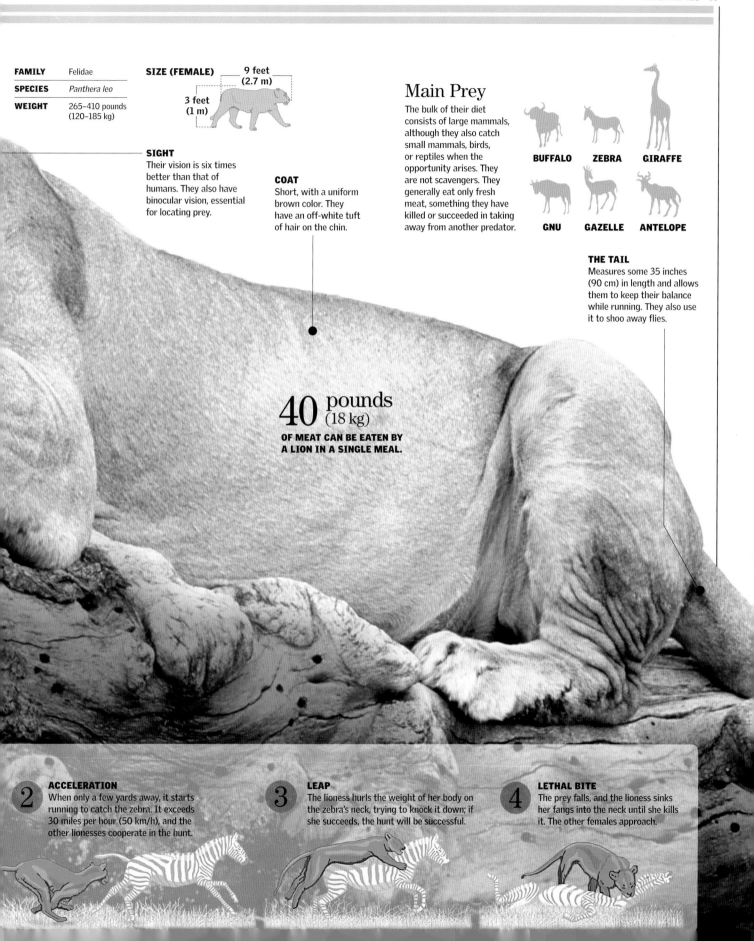

FAMILY	Felidae
SPECIES	*Panthera leo*
WEIGHT	265–410 pounds (120–185 kg)

SIZE (FEMALE)

9 feet (2.7 m)

3 feet (1 m)

SIGHT
Their vision is six times better than that of humans. They also have binocular vision, essential for locating prey.

COAT
Short, with a uniform brown color. They have an off-white tuft of hair on the chin.

Main Prey
The bulk of their diet consists of large mammals, although they also catch small mammals, birds, or reptiles when the opportunity arises. They are not scavengers. They generally eat only fresh meat, something they have killed or succeeded in taking away from another predator.

BUFFALO   **ZEBRA**   **GIRAFFE**

GNU   **GAZELLE**   **ANTELOPE**

THE TAIL
Measures some 35 inches (90 cm) in length and allows them to keep their balance while running. They also use it to shoo away flies.

40 pounds (18 kg)
OF MEAT CAN BE EATEN BY A LION IN A SINGLE MEAL.

2 ACCELERATION
When only a few yards away, it starts running to catch the zebra. It exceeds 30 miles per hour (50 km/h), and the other lionesses cooperate in the hunt.

3 LEAP
The lioness hurls the weight of her body on the zebra's neck, trying to knock it down; if she succeeds, the hunt will be successful.

4 LETHAL BITE
The prey falls, and the lioness sinks her fangs into the neck until she kills it. The other females approach.

Herbivores

Ruminants, such as cows, sheep, or deer, have stomachs made of four chambers with which they carry out a unique kind of digestion. Because these animals need to eat large quantities of grass in very short times—or else be easy targets for predators—they have developed a digestive system that allows them to swallow food, store it, and then return it to the mouth to chew calmly. When animals carry out this activity, they are said to ruminate.

KEY

- 🟦🟦🟦 INGESTION AND FERMENTATION
- ━━━ RUMINATION
- 🟥🟥🟥 REABSORPTION OF NUTRIENTS
- 🟦🟦🟦 ACID DIGESTION
- 🟦🟦🟦 DIGESTION AND ABSORPTION
- 🟦🟦🟦 FERMENTATION AND DIGESTION

Teeth

Herbivorous animals, such as horses and bovids, have molars with a large flat surface that reduces food to pulp, as well as incisors for cutting grass. Grinding is also done by the molars.

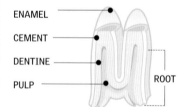

ENAMEL

CEMENT

DENTINE

PULP

ROOT

Cows wrap their tongues around the food.

Then they chew it with lateral movements.

1

Cows lightly chew grass and ingest it into their first two stomachs: the rumen and the reticulum. Food passes continually from the rumen to the reticulum (nearly once every minute). There various bacteria colonies begin fermenting the food.

2

When cows feel satiated, they regurgitate balls of food from the rumen and chew them again in the mouth. This is called rumination; it stimulates salivation, and, as digestion, is a very slow process, cows make use of rumination to improve their own digestion together with the intervention of anaerobic microorganisms, such as protozoa, bacteria, and fungi.

40 gallons (150 L)

OF SALIVA ARE PRODUCED DAILY IN THE PROCESS.

THE RUMINATION PROCESS

helps ruminants reduce the size of the ingested food particles. It is part of the process that allows them to obtain energy from plant cell walls, also called fiber.

Ⓐ REGURGITATION   Ⓑ REMASTICATION   Ⓒ REINSALIVATION   Ⓓ REINGESTION

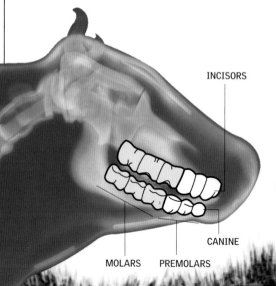

INCISORS

CANINE

MOLARS   PREMOLARS

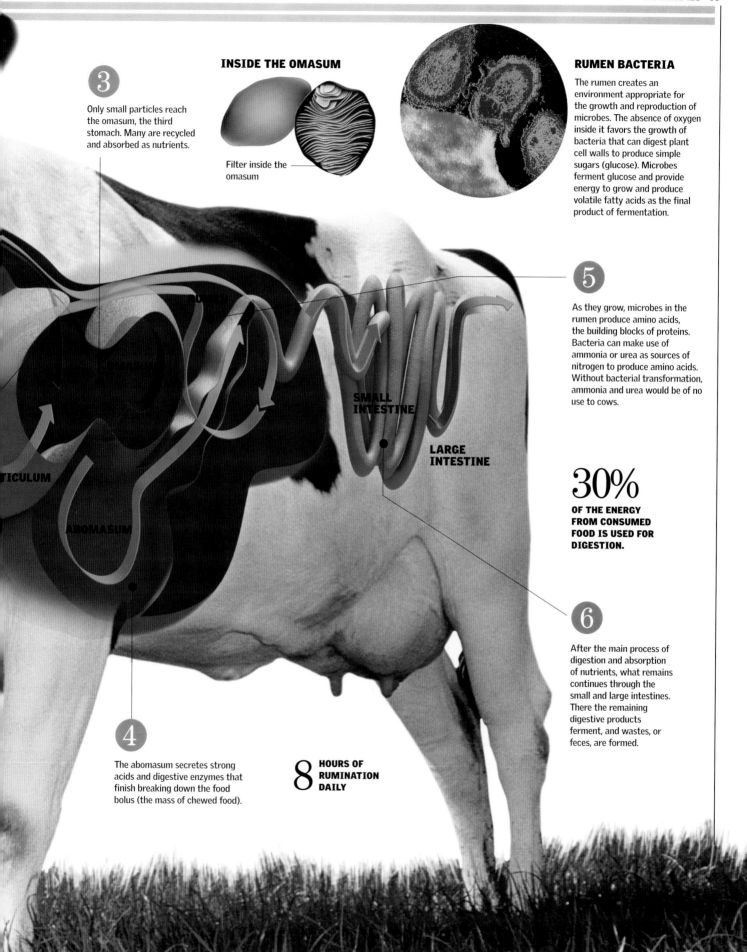

3

Only small particles reach the omasum, the third stomach. Many are recycled and absorbed as nutrients.

INSIDE THE OMASUM

Filter inside the omasum

RUMEN BACTERIA

The rumen creates an environment appropriate for the growth and reproduction of microbes. The absence of oxygen inside it favors the growth of bacteria that can digest plant cell walls to produce simple sugars (glucose). Microbes ferment glucose and provide energy to grow and produce volatile fatty acids as the final product of fermentation.

5

As they grow, microbes in the rumen produce amino acids, the building blocks of proteins. Bacteria can make use of ammonia or urea as sources of nitrogen to produce amino acids. Without bacterial transformation, ammonia and urea would be of no use to cows.

SMALL INTESTINE

LARGE INTESTINE

30%

OF THE ENERGY FROM CONSUMED FOOD IS USED FOR DIGESTION.

TICULUM

ABOMASUM

6

After the main process of digestion and absorption of nutrients, what remains continues through the small and large intestines. There the remaining digestive products ferment, and wastes, or feces, are formed.

4

The abomasum secretes strong acids and digestive enzymes that finish breaking down the food bolus (the mass of chewed food).

8 **HOURS OF RUMINATION DAILY**

The Great Chain

Maintaining ecological balance requires the existence of prey and predators. Predatorial species bring about a sustained reduction in the number of individuals of the prey species. If predators did not exist, their prey would probably proliferate until the ecosystem collapsed, because there would not be enough food for them all. Disappearance of predators is the cause of many imbalances created in certain habitats by humans, whose predatory ability exceeds that of any other living species. Like all other animal species, mammals do not make up a food chain in themselves, instead depending at all times on the participation of plants and other animals.

Level 4

Large carnivores are at the top of the food chain—there are no other predatory species that regulate their population.

SMALL-SPOTTED GENET

Like many highly predatory large felines and dogs, it is in danger of extinction as a result of human activity.

Equilibrium of the System

There is a very efficient natural equilibrium in the food chains of a terrestrial ecosystem, of which mammals form various parts. For this balance to be maintained, there can never be more herbivores than plant food or enough carnivores to overwhelm the herbivores. If there were more herbivores than plant food, they would eat all the vegetation and then suffer a drastic population reduction. A similar situation would occur if there were enough carnivores to overwhelm the herbivores.

Level 3

Small carnivores feed on small, herbivorous mammals or on birds, fish, or invertebrates. At the same time, they must be on guard against other, larger species.

Trophic Pyramid

Energy is transferred from one level to another in an ecosystem. At each level, a small amount of energy is lost. What is retained at one level is the potential energy that will be used by the next. Biomass is the total mass of living matter; it can apply to a specific level of the trophic pyramid, a population of individuals of the same species, or a community of different species.

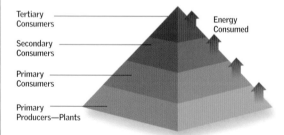

Tertiary Consumers
Secondary Consumers
Primary Consumers
Primary Producers—Plants

Energy Consumed

COMPETITION

Within the same level, different herbivorous rodents (such as rats and prairie dogs) compete with each other for food.

Level 2

Primary consumers devour autotrophic organisms (plants or algae), because they depend on them for subsistence. And other mammals feed on them.

Population

IS GREATER AS ONE GOES DOWN THE PYRAMID.

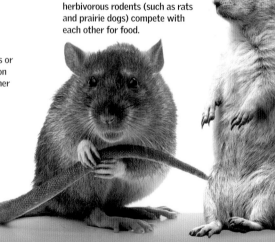

Level 1

Because of photosynthesis, only plants and algae can transform inorganic matter into organic matter. They form the beginning of the food chain.

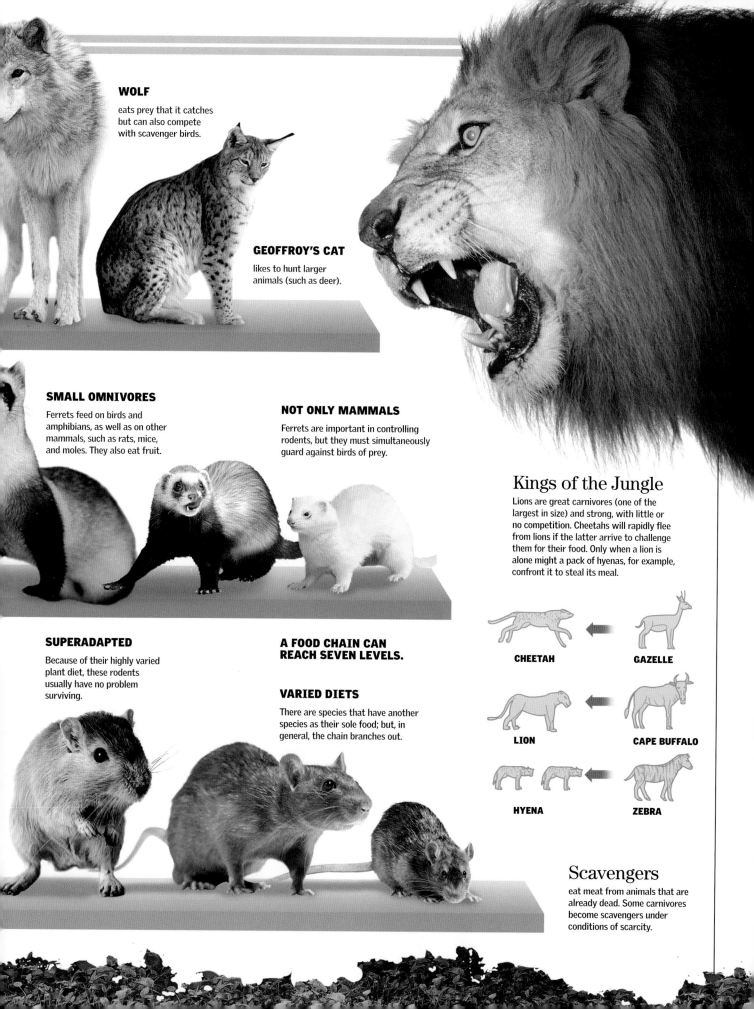

WOLF

eats prey that it catches but can also compete with scavenger birds.

GEOFFROY'S CAT

likes to hunt larger animals (such as deer).

SMALL OMNIVORES

Ferrets feed on birds and amphibians, as well as on other mammals, such as rats, mice, and moles. They also eat fruit.

NOT ONLY MAMMALS

Ferrets are important in controlling rodents, but they must simultaneously guard against birds of prey.

SUPERADAPTED

Because of their highly varied plant diet, these rodents usually have no problem surviving.

A FOOD CHAIN CAN REACH SEVEN LEVELS.

VARIED DIETS

There are species that have another species as their sole food; but, in general, the chain branches out.

Kings of the Jungle

Lions are great carnivores (one of the largest in size) and strong, with little or no competition. Cheetahs will rapidly flee from lions if the latter arrive to challenge them for their food. Only when a lion is alone might a pack of hyenas, for example, confront it to steal its meal.

CHEETAH ← **GAZELLE**

LION ← **CAPE BUFFALO**

HYENA ← **ZEBRA**

Scavengers

eat meat from animals that are already dead. Some carnivores become scavengers under conditions of scarcity.

One for All

Meerkats are small mammals that live in underground colonies, posting guards while the mothers take care of their young. During the day, they go above ground to feed, and at night they go into the burrow to take refuge from the cold. In this large family, made up of dozens of members, each one fulfills a function. When faced with danger, they employ various tactics to defend themselves. One of these is the squeal that lookouts emit in the face of even slight dangers.

MEERKAT
Suricata suricatta

12 inches
(30 cm)

Weight
2 pounds
(1 kg)

FAMILY	Herpestidae
HABITAT	Africa
OFFSPRING	2 to 7

About
30 IS THE NUMBER OF
INDIVIDUALS A
GROUP CAN HAVE.

Social Structure

The social structure is extensive and well defined, ensuring that everyone has a role to fulfill. The lookouts (which may be female or male) take turns to sound the alarm over the arrival of strangers; one that is better fed replaces another that needs to eat. These animals are carnivorous. They eat small mammals, as well as insects and spiders.

FEMALES
must dedicate all their energy to the process of reproducing and feeding and raising young.

OFFSPRING
When the father or mother standing watch gives the cry of danger, all run to hide in the burrow.

BLACK-BACKED JACKAL
The meerkats' largest predator. To detect one before it is seen is of prime importance for the colony.

MARTIAL EAGLES
The most dangerous enemy they have and the one that kills the greatest number of meerkats

Lookout

When a predator is detected, the lookout warns its group so that all of them can take cover in a nearby hole. This role rotates among different members of the group, and the warning is given by a very wide repertoire of sounds, each of which has a distinct meaning.

MEERKATS ALSO USE VOCALIZATIONS TO COMMUNICATE.

Defense

1 SURROUNDING THE ENEMY
They emit a type of squeal. They rock back and forth. They try to appear larger and more ferocious than they are.

2 ON THEIR BACKS
If this tactic fails, they throw themselves down on their backs to protect their necks, showing their fangs and claws.

3 PROTECTION
When it is an aerial predator, they run to hide. If taken by surprise, adults protect the young.

SIGHT
Binocular and in color, it allows them to locate their greatest predators, birds of prey.

HEAD
is kept permanently erect, observing the burrow's surroundings.

VIGILANCE FROM ABOVE
It is common to see them in the highest places of their territory on rocks or tree branches.

FRONT PAWS
They have strong claws, which they use for digging or to defend themselves.

MALES
defend their territory and stand watch. The dominant male is the reproducer.

Territory

The area defended provides the food necessary for the group's subsistence. Males devote themselves to defense, and when resources run out the group migrates to another area.

BURROWS
They dig them with their sharp claws and leave them only during the day.

HIND FEET
They support themselves on their hind feet when they remain standing, keeping watch.

TRIPOD TAIL
Meerkats use it to balance themselves when they are in an upright position.

Diversity

There is great variety among mammals, and in this chapter we try to show you some representatives of the most outstanding differences among them.

For example, here you will discover that there are species, such as bats, that are expert fliers, while others, such as dormice, enter into a deep winter sleep that allows them to save energy

DISTINCTIVE STRIPES
Zebras' stripes extend
down to the underbelly.
They confuse predators.

during times when food is scarce. Here, we will also show you how the bodies of some mammals (whales and dolphins) are adapted to aquatic life. In addition, we will also consider the ability of certain mammals to adapt to the hot and dry conditions of the desert. Camels, in particular, are very adept when it comes to retaining and efficiently using liquids.

Deep Sleep

How many times have you heard the expression "dead as a dormouse"? The comparison is no accident, although it should be understood that dormice do not die: they merely hibernate. In the cold season, low temperatures and scarcity of food lead many mammals to enter into lethargic states. Body temperatures drop, heart rates and respiration slow down, and they lose consciousness.

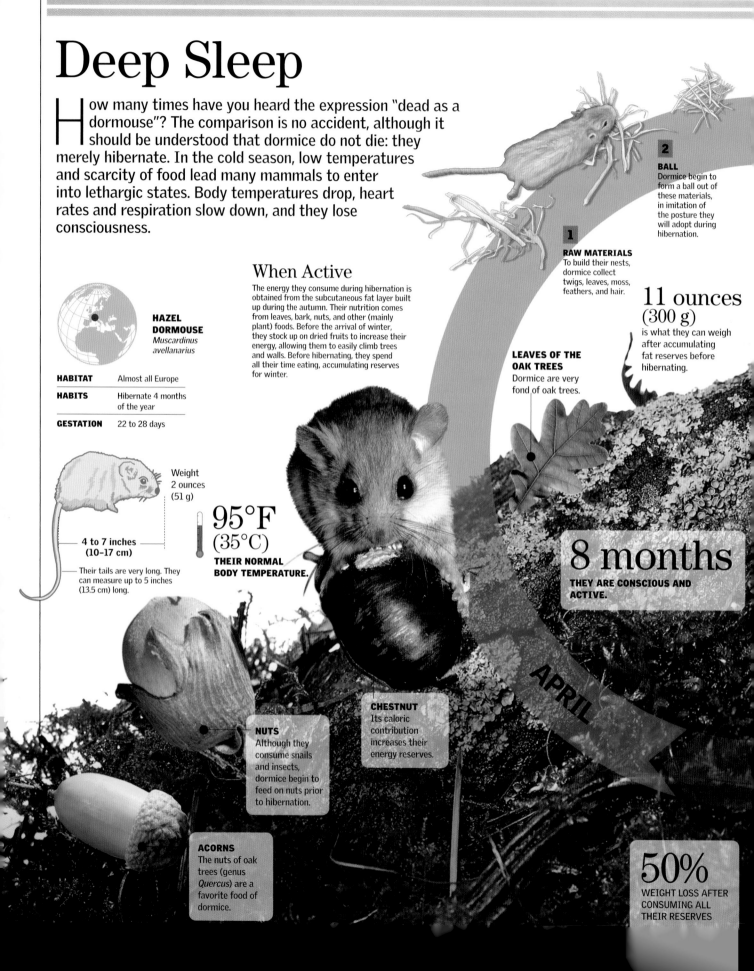

2
BALL
Dormice begin to form a ball out of these materials, in imitation of the posture they will adopt during hibernation.

1
RAW MATERIALS
To build their nests, dormice collect twigs, leaves, moss, feathers, and hair.

When Active

The energy they consume during hibernation is obtained from the subcutaneous fat layer built up during the autumn. Their nutrition comes from leaves, bark, nuts, and other (mainly plant) foods. Before the arrival of winter, they stock up on dried fruits to increase their energy, allowing them to easily climb trees and walls. Before hibernating, they spend all their time eating, accumulating reserves for winter.

HAZEL DORMOUSE
Muscardinus avellanarius

HABITAT	Almost all Europe
HABITS	Hibernate 4 months of the year
GESTATION	22 to 28 days

Weight
2 ounces
(51 g)

**4 to 7 inches
(10–17 cm)**

Their tails are very long. They can measure up to 5 inches (13.5 cm) long.

**95°F
(35°C)
THEIR NORMAL
BODY TEMPERATURE.**

LEAVES OF THE OAK TREES
Dormice are very fond of oak trees.

**11 ounces
(300 g)**
is what they can weigh after accumulating fat reserves before hibernating.

**8 months
THEY ARE CONSCIOUS AND ACTIVE.**

APRIL

CHESTNUT
Its caloric contribution increases their energy reserves.

NUTS
Although they consume snails and insects, dormice begin to feed on nuts prior to hibernation.

ACORNS
The nuts of oak trees (genus *Quercus*) are a favorite food of dormice.

**50%
WEIGHT LOSS AFTER CONSUMING ALL THEIR RESERVES**

Building the Nest

Dormice build their nests out of twigs, moss, and leaves, although they can also hibernate in trees, stone walls, or old buildings, creating a nest from fur, feathers, and leaves. They then settle into the nest, forming a ball. When they cannot find a natural refuge, dormice may settle into birds' nests with total impunity.

3
HOLLOW BALL
Like an ovenbird nest, the ball must be hollow so it can shelter the dormouse.

4
FINISHED NEST
With an entrance in front, the hollow ball has been transformed into a nest.

NOVEMBER

4 months
THEY REMAIN IN A STATE OF HIBERNATION.

Hibernation

During this period, dormice enter into a deep sleep. Body temperature drops to 34°F (1°C), appreciably decreasing the heart rate. In fact, up to 50 minutes can transpire between breaths. Throughout these months, they slowly use up their reserves, losing up to 50 percent of their body weight. Their endocrine system is almost totally at rest: the thyroid ceases functioning, as does the interstitial tissue of the testicles.

34°F (1°C)
THEIR BODY TEMPERATURE DURING HIBERNATION

POSITION OF THE BODY

TAIL
They cover part of the body with it.

HEAD
They hide it behind their long tail.

FEET
remain flexed during these months.

RESPIRATION
Fifty minutes can pass between breaths.

ENERGY
They obtain it from the subcutaneous fat reserves they accumulated in the fall.

HEART
Heartbeats decrease considerably.

BIORHYTHM OF A DORMOUSE WHILE HIBERNATING

TEMPERATURE

WEIGHT

RESPIRATION

Prior Feeding | Deep Hibernation | Brief Activity | Deep Hibernation | After Hibernation

OTHER PLACES FOR HIBERNATION

BIRD'S NEST
If they do not find a place to build their nest, they may take over a bird's nest.

HOLE IN A TREE
can also serve as a burrow for hibernation.

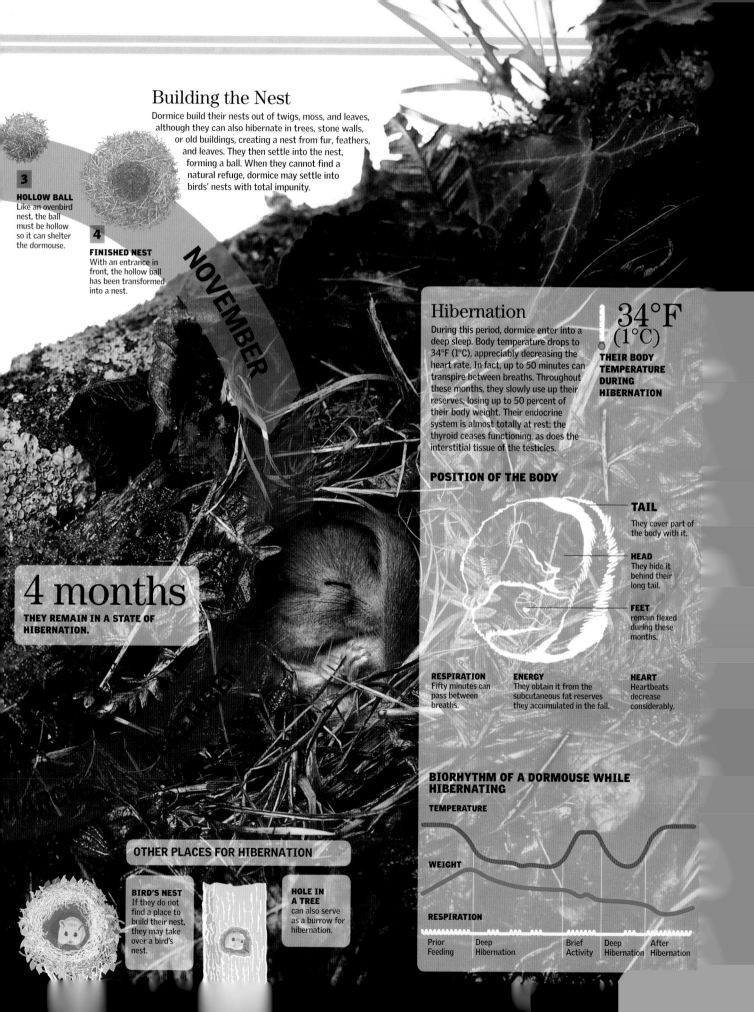

Record Breath-Holders

Sperm whales are unique animals whose species is remarkable for many reasons. On the one hand, they have the ability to dive to a maximum depth of 9,800 feet (3,000 m) and remain underwater without oxygen for up to two hours. They are able to do this by means of a complex physiological mechanism that, for example, can decrease their heart rate, store and use air in the muscles, and prioritize the delivery of oxygen to certain vital organs, such as the heart and lungs. They are the largest whales with teeth, which are found only on the lower mandible.

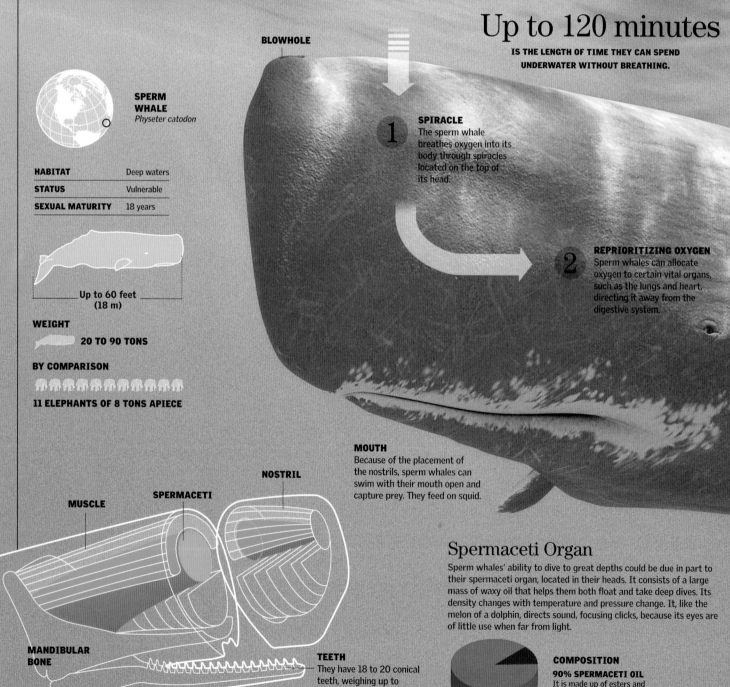

BLOWHOLE

Up to 120 minutes

IS THE LENGTH OF TIME THEY CAN SPEND UNDERWATER WITHOUT BREATHING.

SPERM WHALE
Physeter catodon

HABITAT	Deep waters
STATUS	Vulnerable
SEXUAL MATURITY	18 years

Up to 60 feet (18 m)

WEIGHT

20 TO 90 TONS

BY COMPARISON

11 ELEPHANTS OF 8 TONS APIECE

1 SPIRACLE
The sperm whale breathes oxygen into its body through spiracles located on the top of its head.

2 REPRIORITIZING OXYGEN
Sperm whales can allocate oxygen to certain vital organs, such as the lungs and heart, directing it away from the digestive system.

MOUTH
Because of the placement of the nostrils, sperm whales can swim with their mouth open and capture prey. They feed on squid.

MUSCLE

SPERMACETI

NOSTRIL

MANDIBULAR BONE

TEETH
They have 18 to 20 conical teeth, weighing up to 2 pounds (1 kg) apiece, in each lower mandible.

Spermaceti Organ

Sperm whales' ability to dive to great depths could be due in part to their spermaceti organ, located in their heads. It consists of a large mass of waxy oil that helps them both float and take deep dives. Its density changes with temperature and pressure change. It, like the melon of a dolphin, directs sound, focusing clicks, because its eyes are of little use when far from light.

COMPOSITION
90% SPERMACETI OIL
It is made up of esters and triglycerides.

Adaptation in Respiration

When they dive to great depths, sperm whales activate an entire physiological mechanism that makes maximum use of their oxygen reserves. This produces what is called a thoracic and pulmonary collapse, causing air to pass from the lungs to the trachea, reducing the absorption of the toxin nitrogen. They also rapidly transmit nitrogen from the blood to the lungs at the end of the dive, thus reducing the circulation of blood to the muscles. Sperm whales' muscles contain a large amount of myoglobin, a protein that stores oxygen, allowing the whales to stay underwater much longer.

BLOWHOLE
Upon submerging, it fills with water, which cools the spermaceti oil and makes it denser.

HEART
The heart rate slows down during the dive, limiting oxygen consumption.

BLOOD
An ample blood flow, rich in hemoglobin, transports elevated levels of oxygen to the body and brain.

ON THE SURFACE
Blowhole remains open, allowing the whales to breathe as much oxygen as they can before diving.

WHEN THEY DIVE
powerful muscles tightly close the opening of the blowhole, keeping water from entering.

RETIA MIRABILIA
The retia is a network of blood vessels (mirabilia) that filter the blood entering the brain.

LUNGS
absorb oxygen very efficiently.

3 BRADYCARDIA
During a dive, the heart rate drops (a condition known as bradycardia), which lowers oxygen consumption.

TAIL
is large and horizontal and is the whale's main means of propulsion.

Dive

True diving champions, sperm whales can dive to depths of 9,800 feet (3,000 m), descending up to 10 feet (3 m) per second in search of squid. As a general rule, their dives last about 50 minutes, but they can remain underwater up to two hours. Before beginning a deep dive, they lift their caudal fin completely out of the water. They do not have a dorsal fin, but they do have a few triangular humps on the posterior part of their body.

0 FEET (0 M) ON THE SURFACE
They inhale oxygen through the blowhole located at the top of the head.

+ 3,300 FEET (1,000 M) 90 MINUTES
They store 90 percent of their oxygen in their muscles, so they can be submerged for a long time.

0 FEET (0 M) ON THE SURFACE
They exhale all the air from their lungs; this is called spouting, or blowing.

Making Use of Oxygen

Sperm whales can dive deeper and stay submerged longer than any other mammal, because they have various ways of saving oxygen: an ability to store it in their muscles, a metabolism that can function anaerobically, and the inducement of bradycardia during a dive.

15% AMOUNT OF AIR REPLACED IN ONE BREATH

85% AMOUNT OF AIR REPLACED IN ONE BREATH

Aerial Acrobatics

Cats have a surprising ability to land upright. The secret lies in their skeleton, which is more flexible and has more bones than that of any other mammal. Cats' reflexes allow them to twist using the physical principle of the conservation of angular momentum. The principle, first formulated by Isaac Newton, states that all bodies in circular movement tend to use a constant amount of energy. Thus, the more the animal extends its legs to its axis of rotation, the slower it rotates, redistributing the total energy of the system. If the animal tucks in its legs, it rotates more rapidly.

NAME	Domestic cat
FAMILY	Felidae
SPECIES	Felis catus
ADULT WEIGHT	4 to 15 pounds (2–7 kg)
LONGEVITY	15 years
DIMENSIONS	

10 INCHES (25 CM)

4 INCHES (10 CM)

12 INCHES (30 CM)

FORCE OF GRAVITY

AXIS

THE "ACCELERATOR"
The cat folds its front legs into its axis to increase the speed of rotation of this part. It rotates 180 degrees.

THE "BRAKE"
It extends its hind legs perpendicular to the axis and reduces the speed of rotation of this part.

It extends its front legs at right angles to the axis.

It draws its hind legs into the axis of the body.

1 STARTS UPSIDE DOWN
The cat begins to fall upside down and will turn 180 degrees upon its axis (in two stages), landing upright.

2 FIRST TWIST
In this maneuver, the cat rotates the front half of its body 180 degrees on its body's axis. The other half rotates only slightly as a result.

AXIS

SLIGHT ROTATION

BACK HALF

STRONG ROTATION

FRONT HALF

3 WITH INDEPENDENCE
Like a skater who extends or folds the arms to control the speed of rotation, the cat moves its hind legs—but independently of each other.

LIKE A SKATER

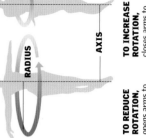

RADIUS

AXIS

TO REDUCE ROTATION, opens arms to increase the radius of rotation.

TO INCREASE ROTATION, closes arms to reduce the radius of the rotation.

Strategy

Although it seems contradictory, it has been shown that a cat can suffer greater injury the shorter the distance it falls. When a cat senses that it is falling, it compacts its body and stretches out its legs so that when it lands on the ground it can soften the impact. If the fall is from less than one story in height, the cat does not have time to assume this posture.

HARM

HEIGHT

○ TERMINAL VELOCITY

RELAXATION

DEFENSIVE POSTURE

FIRST TWIST

4 SECOND TWIST

The cat lowers its hind legs and completes a full rotation on its axis. It again carries out two more rotations, one tighter than the other:

AXIS

STRONG ROTATION

BACK HALF

SLIGHT ROTATION

FRONT HALF

BACK HALF
Now the folded legs increase the speed of rotation of this part.

FRONT HALF
The extended legs reduce the speed of rotation of this part. It rotates 180 degrees.

The tail stabilizes the weight of the body during the descent.

5 FOUR FEET PLACED UNDER THE BODY

With four feet positioned under the body, the cat bends its spine like a parachute and then merely corrects its posture for landing.

It extends the hind legs to the height of the front legs.

11% ELONGATION CAPACITY

EXTREME FLEXIBILITY
Cats do not have a clavicle, and the articulations of their vertebrae are more flexible than those of most mammals. They can travel five times the length of their body in one leap.

1/8 of a second
TIME IT TAKES TO ROTATE AND LAND ON ITS FEET 1/2 SECOND LATER

6 LANDING

Its front legs make the first contact with the ground. Then it lands on its hind legs, and, finally, it relaxes its tail.

At the moment of landing, the cat slightly flexes its feet to cushion the blow.

Equilibrium

The inner ear in the temporal bone is divided into the cochlea, the vestibule, and three semicircular canals. Inside, there is a system of cilia (sense receptors) and a viscous substance (endolymph) that generates the sense of balance when the two come in contact with each other.

CROSS SECTION OF A SEMICIRCULAR CANAL

BULLA
It holds the cilia, which are equilibrium receptors.

During a rotation, endolymph moves the cilia in the direction opposite the body's motion.

QUICK AND PRECISE SHAKE
During the rotation, endolymph can splash into the semicircular canals. To return the liquid to its place, the cat gives a quick shake of its head.

INNER EAR

COCHLEA

Natural Builders

They have no bricks or cement, but beavers, semiaquatic rodents, skillfully manage to build lodges of great architectural beauty. They do not work alone, and it is usual for them to act in family groups. Everyone collaborates in building the home, which is generally located next to a river or lake surrounded by forested areas and which can be entered only through aquatic tunnels. The task is difficult, and beavers work their whole lives enlarging, repairing, and improving their dwelling.

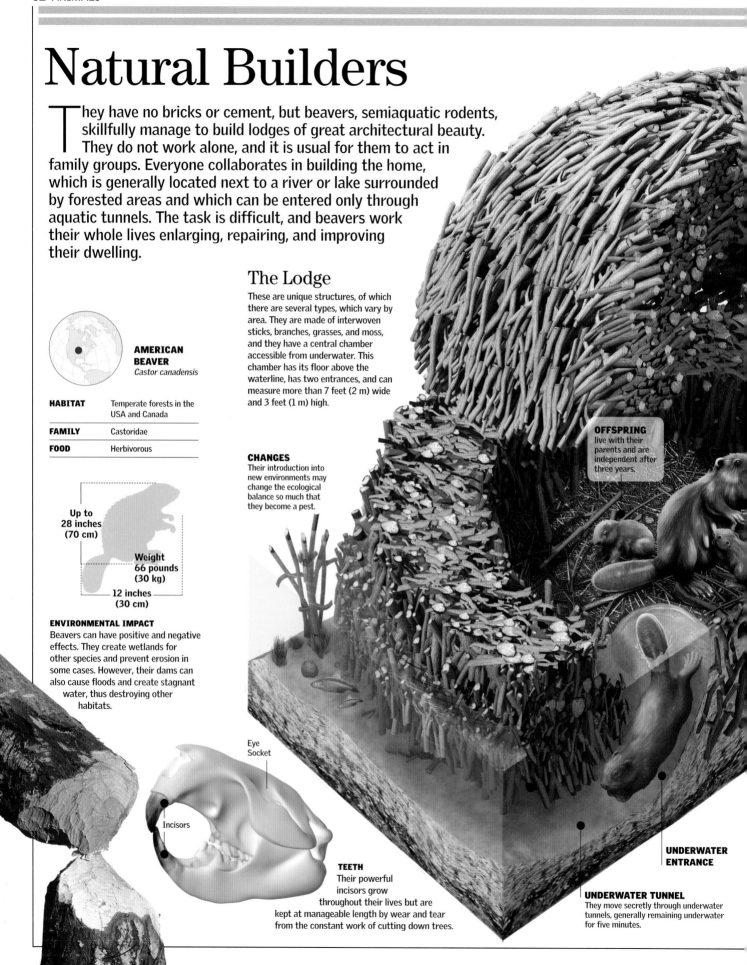

AMERICAN BEAVER
Castor canadensis

HABITAT	Temperate forests in the USA and Canada
FAMILY	Castoridae
FOOD	Herbivorous

Up to 28 inches (70 cm)

Weight 66 pounds (30 kg)

12 inches (30 cm)

ENVIRONMENTAL IMPACT
Beavers can have positive and negative effects. They create wetlands for other species and prevent erosion in some cases. However, their dams can also cause floods and create stagnant water, thus destroying other habitats.

The Lodge

These are unique structures, of which there are several types, which vary by area. They are made of interwoven sticks, branches, grasses, and moss, and they have a central chamber accessible from underwater. This chamber has its floor above the waterline, has two entrances, and can measure more than 7 feet (2 m) wide and 3 feet (1 m) high.

CHANGES
Their introduction into new environments may change the ecological balance so much that they become a pest.

OFFSPRING
live with their parents and are independent after three years.

Eye Socket

Incisors

TEETH
Their powerful incisors grow throughout their lives but are kept at manageable length by wear and tear from the constant work of cutting down trees.

UNDERWATER ENTRANCE

UNDERWATER TUNNEL
They move secretly through underwater tunnels, generally remaining underwater for five minutes.

2 times

THE STRENGTH OF THEIR INCISOR TEETH (USED FOR CHEWING) IN COMPARISON WITH HUMANS

Dam

Beavers continually repair the dam and add materials to it. Floating material carried along by the water is retained in the dam, along with the roots of vegetation that grows upon it, strengthening the entire structure.

Lodge

Dry Area

Water Level

Underwater Entrance

Dam

ROOF
Made of trunks, branches, stones, and mud.

THE DAM
has two purposes—first, to raise the water level; and second, to enlarge the flooded area around the den. Dams are built out of sticks and tree trunks.

DRY AREA
Covered with tree bark, grass, and little pieces of wood

EXIT
Beavers have webbed feet that they use to dive and for other quick movements.

ROCKS
maintain the structure of the dam, holding the tree trunks in place.

UNDERWATER ENTRANCE

TECHNIQUES

Beavers frequently work in groups to gnaw down a trunk and carry it away. One of them cuts the tree with its teeth while the others stand guard. This work takes about 15 minutes, and then the tree falls.

Their mandibles and teeth are strong, and they use their front feet as hands.

THE FOUNDATION
In winter, they store fresh branches in the pond to serve as a food reserve.

BRANCHES
The material most used in constructing the lodge. They are used to make the ceiling and to keep the inside dry.

15 minutes
IS HOW LONG A BEAVER CAN STAY UNDERWATER WHEN THREATENED.

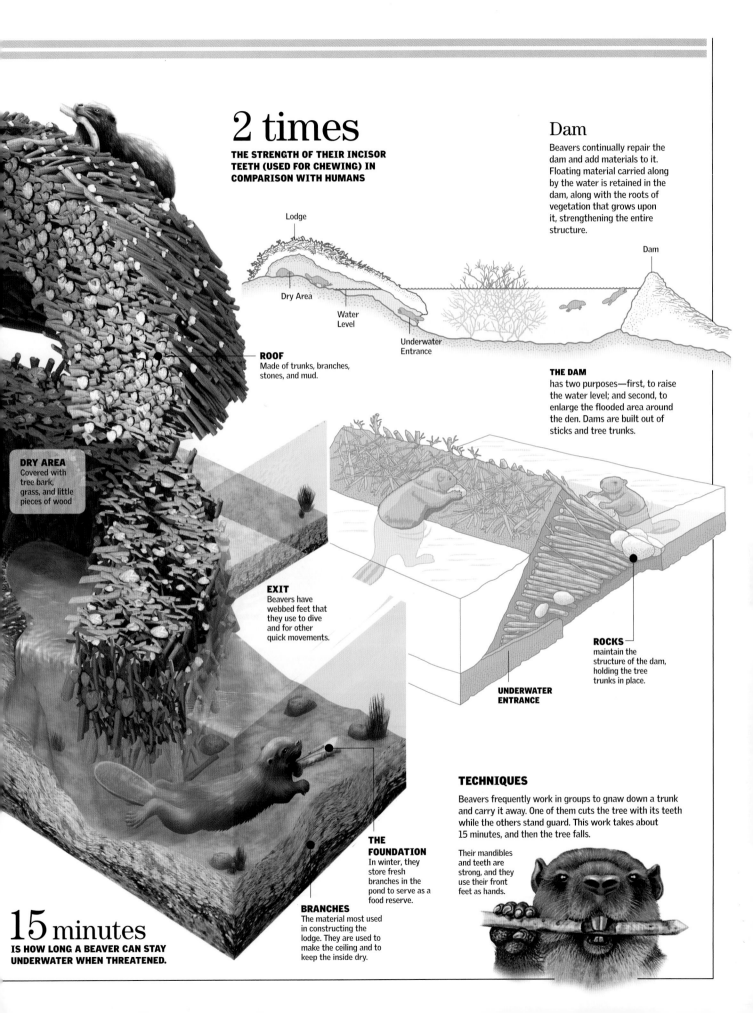

Nocturnal Flight

Bats are the only mammals that can fly. Scientists call them Chiroptera, a term derived from Greek words meaning "winged hands." Their forelimbs have been transformed into hands with very long fingers joined together by a membrane (called the patagium) that forms the surface of the wing. These mammals' senses are so sensitive that they can move and hunt quickly and accurately in the dark.

Expert Pilots

Moved by their chest and back muscles, bats' wings push downward and backward, generating both thrust and lift. Then the wings spread sideways and upward. Finally, they move forward until the tips almost rub the bat's head. Many of these flying mammals can drift through the air, gliding without flapping and maneuvering by folding their wings.

Their Radar

Most of the time bats fly at night in near-total darkness. Instead of light, they use a natural system similar to sonar or radar to guide themselves. This system uses acoustical signals the bats themselves emit while flying. This system allows them to recognize the location of any object in front of them or of prey, along with its direction, size, or speed. It is as if they were seeing without light.

1 The animal emits an acoustical vibration imperceptible to the human ear because of its high frequency (about 18 kHz). The signal strikes the objects around it.

2 When the signals bounce back, the bat perceives their intensity and phase difference—the faster and more intense the return signal, the nearer the object or prey.

60 miles per hour
(97 km) THE SPEED SOME BATS MAY REACH DURING FLIGHT

Hibernation

These bats spend the winter in a lethargic state hanging by their feet, faces down, in caves and other dark places. Bats are warm-blooded animals while they are active and become similar to cold-blooded creatures when they are asleep. They enter into a state of hibernation more rapidly and easily than any other mammal, and they can survive in cold temperatures for many months—even inside refrigerators—without needing to feed.

FRUIT BAT (FRANQUET'S EPAULETTED BAT)
Epomops franqueti

HABITAT	Forests of Ghana and Congo
FAMILY	Pteropodae
LENGTH OF WINGSPAN	14 inches (36 cm)

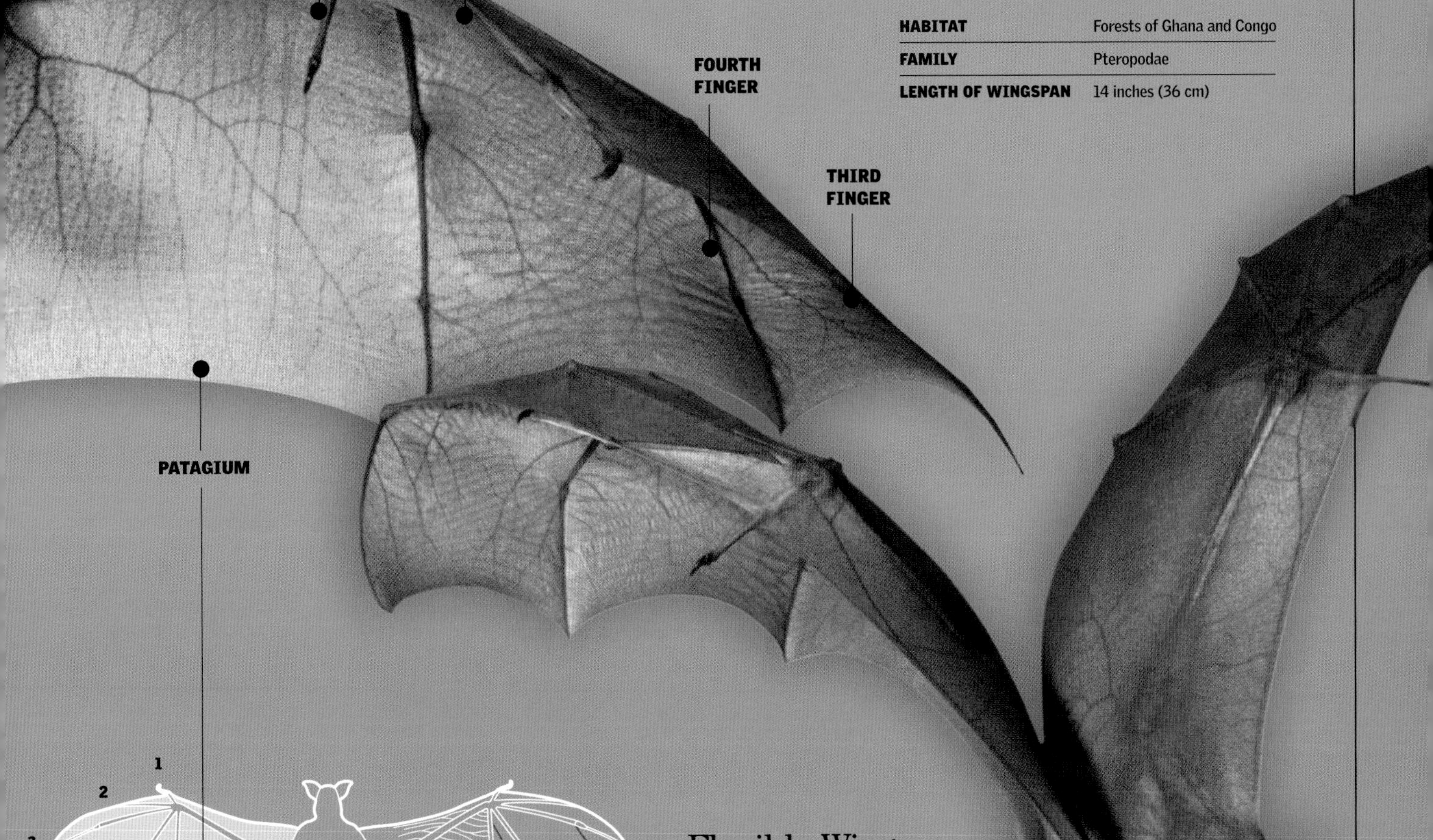

HUMERUS

RADIUS

THUMB

SECOND FINGER

FOURTH FINGER

THIRD FINGER

PATAGIUM

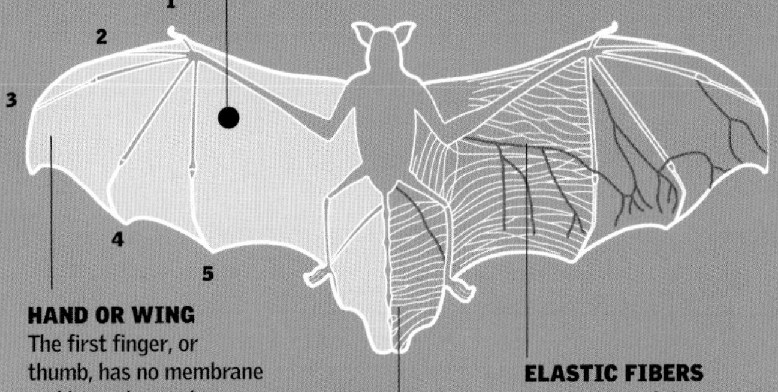

1
2
3
4
5

HAND OR WING
The first finger, or thumb, has no membrane and is used as a claw. Powerful muscles move the entire wing.

UROPATAGIUM

ELASTIC FIBERS
The texture of the wing is soft and flexible. It is lined with blood vessels.

Flexible Wings

The patagium is formed by the membranes between the digits. In some species, the wings are also extended by an additional membrane (uropatagium), which joins the hind limbs to the tail. Their wings are not only used for flying (pushing the air as if they were oars in water) but also help to maintain a constant body temperature and to trap insects, upon which bats feed.

Playing Hide and Seek

Just like other species of the animal kingdom, some mammals that live in the wild rely on their body's colorations or appearances to disguise their presence. Some mammals imitate objects in their environment, and others take on the appearances of other animals. Zebras' stripes, for example, give these animals a very showy appearance—but when moving in their natural environment, zebras are camouflaged. Some differentiate between mimicry and crypsis, which is the natural ability to go unnoticed without requiring any associated behavior. In other cases, however, the forms and colors of camouflage would be useless if they were not accompanied by some kind of imitative behavior

Evolutionary Adaptations

Mimicry is defined as the ability of some living beings to imitate the appearance of another living being or an inanimate object in the environment. Protective mimicry is the camouflage used by animals incapable of defending themselves in any other way. Aggressive mimicry, on the other hand, allows organisms to surprise and attack their prey. This occurs, for example, with wild felines (mountain lions, ocelots, lynx), which take advantage of their skin colors and the patterns of their fur to go unnoticed in their ecosystems. Zebras travel in herds as a natural form of self-protection. The disruptive coloration of their coats makes it difficult for predators that rely on speed and sharp senses to distinguish one individual prey from another. Kicking and biting, zebras collectively defend themselves from attacks by feline predators. These felines also make use of camouflage strategies to make their attacks one on one. Many animals use elements from their surroundings or even of other living organisms to camouflage themselves. Sloths are another example; being the slowest of the mammals, they have no choice but to cover themselves in algae to avoid notice.

STRIPES
The coloration of their coat changes with the incidence and intensity of sunlight.

SPOTS
allow giraffes to conceal themselves among the high leaves they reach with their long neck.

PATTERNS
are irregular forms between stripes that allow tigers to lie in ambush for their prey among thickets.

Different Patterns

The pattern of a zebra's coat does not exactly copy the shapes and colors of objects in the wild environment surrounding it. Nevertheless, it does have patterns that allow it, with the help of certain behaviors and motions, to disguise its appearance in more than one setting of the zebra's natural habitat. In the case of Arctic animals, it is the uniform white color of the winter environment that determines the way in which species camouflage themselves.

In Motion

The patterns of tigers' coats are useful in concealing their contours, especially when they are moving among the shrubs and bushes of the plains where they hunt. Elk horns, however, can be concealed among the vegetation they resemble for only as long as they keep still.

Disruptive Coloration

The body's contours are blurred when some spots of color are much darker or lighter than the rest of the coat.

Part of the Hideaway

Chipmunks (*Tamias* species) live in coniferous or deciduous forests, where they feed on nuts, insects, eggs, seeds, and other plant foods. The colors of their coats are essential, because—although they are very skillful at moving in the upper branches—their small size and short legs make them very vulnerable when they are on the ground.

PROTECTIVE SURROUNDINGS
Many have a coat that changes color depending on the surroundings.

FUR
Shades and differences of color in the coat are similar to those of tree trunks and dry leaves.

The Language of Water

The ways in which cetaceans communicate with others of their kind are among the most sophisticated in the animal kingdom. Dolphins, for example, click with their mandibles when in trouble and whistle repeatedly when afraid or excited. During courtship and mating, they touch and caress. They also communicate through visual signals—such as leaping—to show that food is close by. They have a wide variety of ways to transmit important information.

HAVING FUN

Play for dolphins, as with other mammals, fulfills an essential role in the formation of social strata.

MELON

is an organ filled with low-density lipids that concentrate and direct the pulses emitted, sending waves forward. The shape of the melon can be varied to better focus the sounds.

COMMON NAME	Bottlenose dolphin
FAMILY	Delphinidae
SPECIES	*Tursiops truncatus*
ADULT WEIGHT	330 to 1,400 pounds (150 to 650 kg)
LONGEVITY	30 to 40 years

7 to 13 feet (2-4 m)

THEY REACH 22 MPH (35 KM/H)

SPIRACLE   LIP

NASAL AIR SAC

DORSAL FIN
allows dolphins to maintain their equilibrium in the water.

INHALATION
1 The spiracle opens so oxygen can enter.

SPIRACLE

Air to the lungs

2 The nasal air sacs begin to inflate.

They can go for 12 minutes without taking in oxygen.

LARYNX

HOW THE SOUND IS PRODUCED

4 The nasal air sacs deflate

Melon

Air in the lungs

CAUDAL FIN
has a horizontal axis (unlike that of fish), which serves to propel dolphins forward.

SOUND

1 **Emission**

Sounds are generated by air passing through the respiratory chambers. But it is in the melon that resonance is generated and amplified. Greater frequencies and intensities are achieved in this way.

3 **EXHALATION**
Air resonates in the nasal sacs and the produced sound is directed through the melon.

Brain

PECTORAL FIN

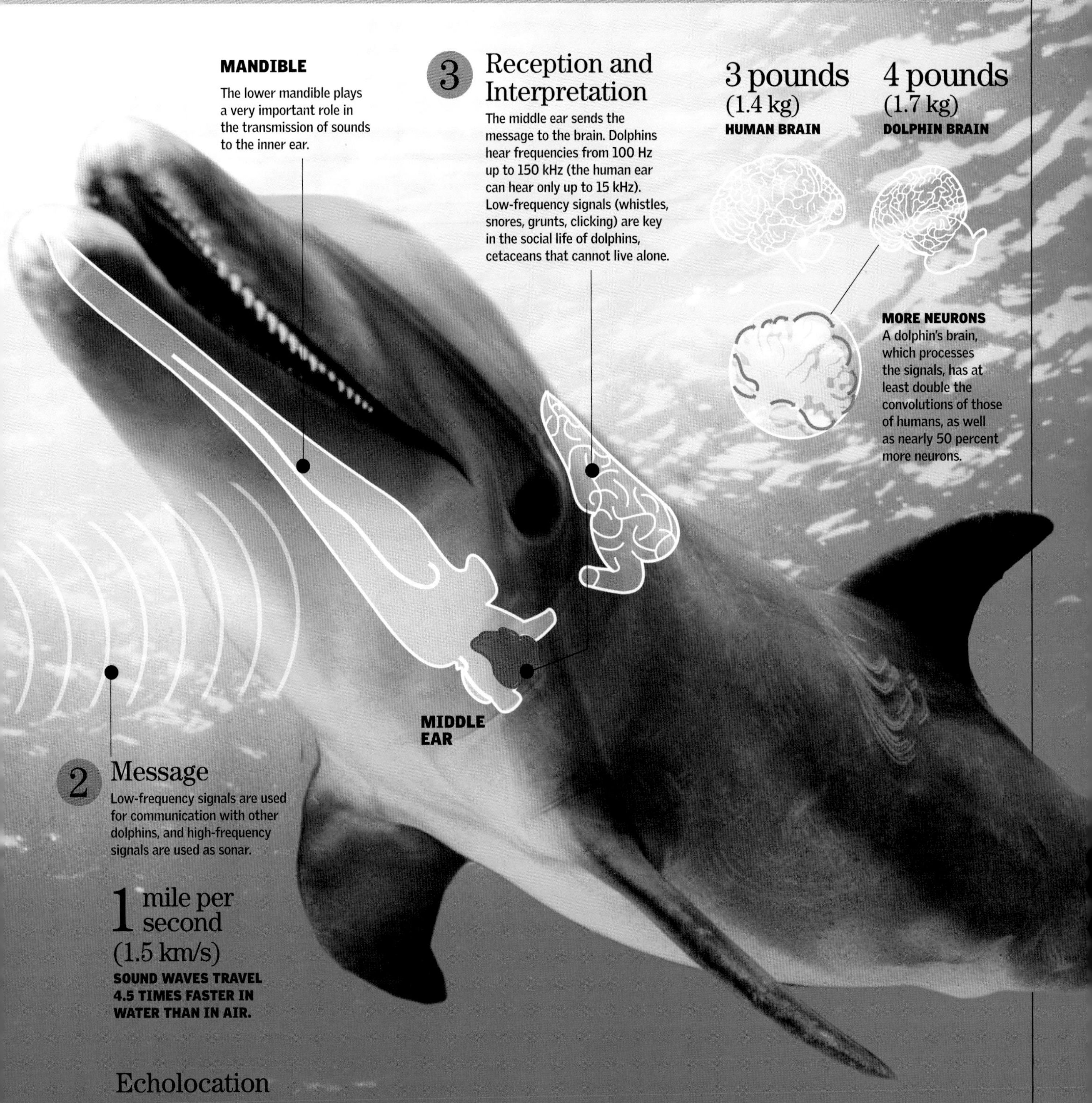

MANDIBLE

The lower mandible plays a very important role in the transmission of sounds to the inner ear.

3 Reception and Interpretation

The middle ear sends the message to the brain. Dolphins hear frequencies from 100 Hz up to 150 kHz (the human ear can hear only up to 15 kHz). Low-frequency signals (whistles, snores, grunts, clicking) are key in the social life of dolphins, cetaceans that cannot live alone.

3 pounds
(1.4 kg)
HUMAN BRAIN

4 pounds
(1.7 kg)
DOLPHIN BRAIN

MORE NEURONS

A dolphin's brain, which processes the signals, has at least double the convolutions of those of humans, as well as nearly 50 percent more neurons.

MIDDLE EAR

2 Message

Low-frequency signals are used for communication with other dolphins, and high-frequency signals are used as sonar.

1 mile per second
(1.5 km/s)
SOUND WAVES TRAVEL 4.5 TIMES FASTER IN WATER THAN IN AIR.

Echolocation

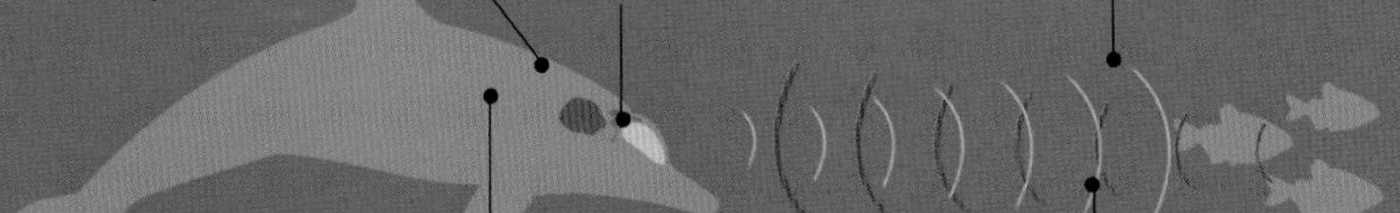

A The dolphin emits a series of clicking sounds from the nasal cavity.

B The melon concentrates the clicks and projects them forward.

C These waves bounce off objects they encounter in their way.

E The intensity, pitch, and return time of the echo indicate the size, position, and direction of the obstacle.

D Part of the signal bounces back and returns to the dolphin in the form of an echo.

SIGNAL WITH ECHO

CLICK   CLICK

ECHO   ECHO

0 S   6 S   12 S   18 S

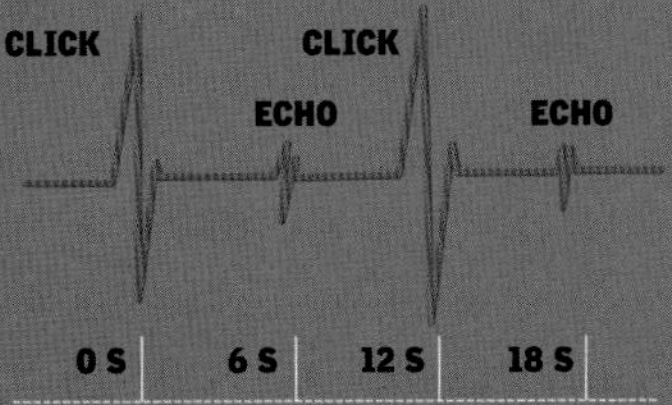

2 BIRDS

The Nature of Birds

Many scientists maintain that birds descended from dinosaurs because fossils of dinosaur specimens with feathers have been found. As a group, birds have exceptional eyesight—they have the largest eyes in relation to the size of their bodies. In addition, they have very light bones, which are suitable for flight. Just like

BIRDS **73**

OWL
(Bubo capensis)
This owl is native to
Africa. It feeds on
birds and mammals.

74 Skeleton And Musculature
76 Internal Organs
78 The Senses
80 Feathers
82 Wings To Fly
84 Types Of Tail
86 Gliding
88 Flapping Flight

their bills, birds' feet have also changed in accordance with the functions and particular needs of each species. For instance, walking birds—like other vertebrate groups—display a marked tendency toward having a reduced number of toes; ostriches, for example, have only two. Some birds of prey, such as eagles, have feet that are veritable hooks.

Skeleton and Musculature

Both lightweight and resistant, the skeleton of birds underwent important changes in order to adapt to flight. Some bones, like those of the skull and wings, fused to become lighter. Birds have fewer bones than other vertebrates. Because their bones are hollow, containing internal air chambers, the total weight of their bones is less than that of their feathers. Birds' spines tend to be very flexible in the cervical region and rigid near the rib cage, where a large, curved frontal bone called the sternum attaches. The sternum features a large keel, to which the pectoral muscles attach. These large, strong muscles are used for flapping the wings. In contrast, running birds, such as ostriches, have more developed muscles in their legs.

EYE SOCK

Flapping Wings

Flying demands an enormous amount of energy and strength. Consequently, the muscles responsible for flapping the wings become very large, easily comprising 15 percent of the weight of a flying bird. Two pairs of pectorals, in which one muscle of the pair is bigger than the other, work to raise and lower the wings. They function symmetrically and in opposition to each other: when one contracts, the other relaxes. Their placement within the thoracic cavity corresponds roughly to the bird's center of gravity. The motion of the wings also requires strong tendons.

HUMMINGBIRD
Because of its adaptation to stationary flight, its pectoral muscles can account for 40 percent of its total weight.

SKULL
Light because of the fusing of bones, the skull does not have teeth, a bony jaw, or grinding muscles.

UPPER MANDIBLE OF BILL
In some species, it is flexible.

LOWER MANDIBLE OF BILL
It is flexible, allowing birds to open their mouths wide.

FURCULA (COLLARBONE)
Known as the wishbone, it is unique to birds and results from the fusion of the collarbones.

STERNUM
Hyperdeveloped in flying birds, the sternum's long keel facilitates the attachment of the pectorals.

WINGS

Without a doubt, wings are the greatest adaptation of birds. Strong tendons travel through the wings and merge into the hand bones, where the feathers are attached.

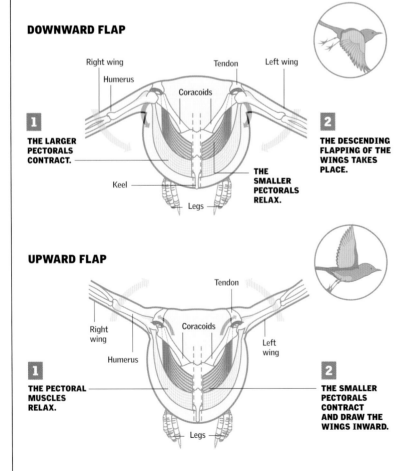

DOWNWARD FLAP

Right wing
Humerus
Coracoids
Tendon
Left wing

1 THE LARGER PECTORALS CONTRACT.

Keel
Legs

THE SMALLER PECTORALS RELAX.

2 THE DESCENDING FLAPPING OF THE WINGS TAKES PLACE.

UPWARD FLAP

Tendon
Right wing
Coracoids
Left wing
Humerus

1 THE PECTORAL MUSCLES RELAX.

Legs

2 THE SMALLER PECTORALS CONTRACT AND DRAW THE WINGS INWARD.

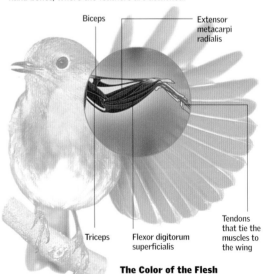

Biceps
Extensor metacarpi radialis
Triceps
Flexor digitorum superficialis
Tendons that tie the muscles to the wing

The Color of the Flesh
depends on the blood circulation in the muscles: the more circulation, the redder the flesh. Flying birds have red flesh, whereas nonflying birds, such as chickens, have white flesh.

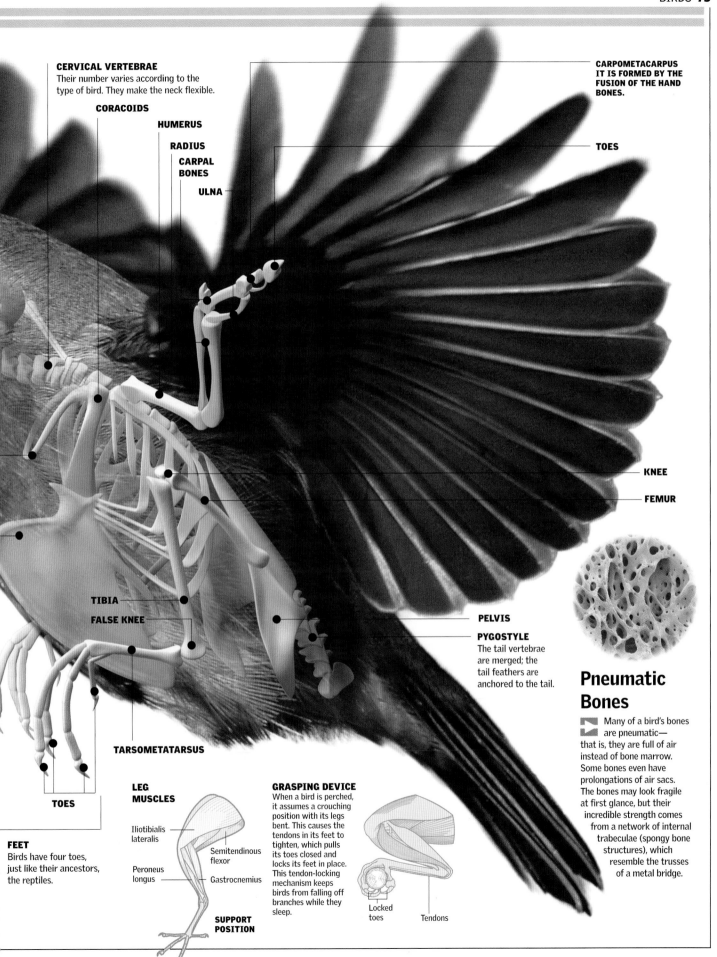

CERVICAL VERTEBRAE
Their number varies according to the type of bird. They make the neck flexible.

CORACOIDS

HUMERUS

RADIUS

CARPAL BONES

ULNA

CARPOMETACARPUS
IT IS FORMED BY THE FUSION OF THE HAND BONES.

TOES

KNEE

FEMUR

TIBIA

FALSE KNEE

PELVIS

PYGOSTYLE
The tail vertebrae are merged; the tail feathers are anchored to the tail.

TARSOMETATARSUS

TOES

FEET
Birds have four toes, just like their ancestors, the reptiles.

LEG MUSCLES

Iliotibialis lateralis

Semitendinous flexor

Peroneus longus

Gastrocnemius

SUPPORT POSITION

GRASPING DEVICE
When a bird is perched, it assumes a crouching position with its legs bent. This causes the tendons in its feet to tighten, which pulls its toes closed and locks its feet in place. This tendon-locking mechanism keeps birds from falling off branches while they sleep.

Locked toes

Tendons

Pneumatic Bones

Many of a bird's bones are pneumatic—that is, they are full of air instead of bone marrow. Some bones even have prolongations of air sacs. The bones may look fragile at first glance, but their incredible strength comes from a network of internal trabeculae (spongy bone structures), which resemble the trusses of a metal bridge.

Internal Organs

Birds in flight can consume oxygen at a rate that a well-trained athlete would not be able to withstand for even a few minutes. Because of this oxygen consumption, all their organs have had to adapt. The lungs of birds, though smaller than those of mammals of similar size, are much more efficient. Their lungs have several air sacs that both increase the efficiency of their respiratory systems and make them lighter. A special feature of the digestive system is a crop in the esophagus, where food is stored for digestion or for feeding the young. A bird's heart can be four times larger in relation to its body size than a human's in relation to its body size.

Rufous Hummingbird
(Selasphorus rufus)

Digestive System

Birds have no teeth. They, therefore, ingest food without chewing, and their stomachs break it down. The stomach is divided into two parts: the glandular (or proventriculus) part, which secretes acids, and the muscular (or gizzard) part, whose muscular walls grind up what is eaten. In general, the process is very fast because flying requires a lot of energy, and the bird has to replenish that energy quickly. The digestive system ends at the cloaca, which is an excretory orifice shared with the urinary system. Birds absorb almost all the water they drink.

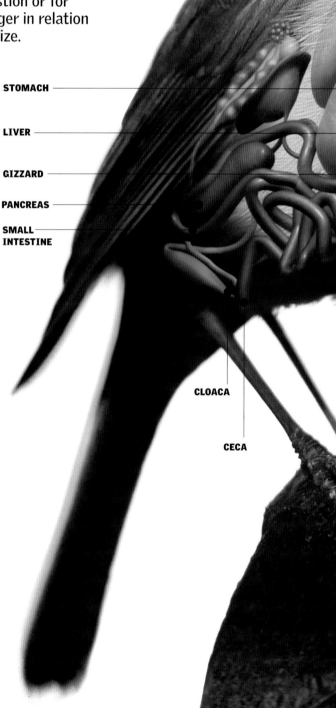

- STOMACH
- LIVER
- GIZZARD
- PANCREAS
- SMALL INTESTINE
- CLOACA
- CECA

FOOD ITINERARY

1 STORAGE
Some birds have a crop, which enables them to store food and digest it later. This way they decrease their exposure to predators.

2 PRODUCTION
The proventriculus secretes the gastric juices that initiate digestion.

3 BREAKDOWN
In the gizzard, a strong and muscular pouch, food is broken down with the help of swallowed stones or sand. The stones and sand play the role of teeth.

4 WATER ABSORPTION
occurs in the small intestine. Birds normally get water from the food they ingest.

5 EXCRETION
The cloaca expels feces mixed with urine coming from the excretory system.

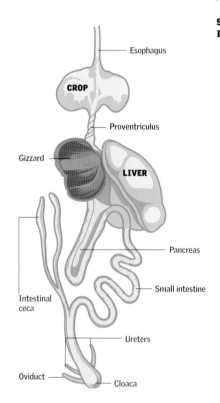

- Esophagus
- CROP
- Proventriculus
- Gizzard
- LIVER
- Pancreas
- Small intestine
- Intestinal ceca
- Ureters
- Oviduct
- Cloaca

TYPES OF GIZZARD

GRANIVOROUS BIRDS
have thick muscle walls and strong mucous membranes (or internal skin) to break down seeds.

CARNIVOROUS BIRDS
have thin muscle walls because digestion takes place in the proventriculus.

TONGUE
Usually short, narrow, triangular, and not very muscular.

ESOPHAGUS

TRACHEA

SYRINX
Makes it possible for birds to sing.

LUNG
Almost rigid because of its structure

CROP

STERNUM

HEART

A HUMMINGBIRD'S HEART BEATS

700 times a minute.

Respiratory System

Birds have the most efficient respiratory system of any vertebrate because of the great effort that flying demands. It has two small, almost rigid lungs that are assisted by nine air sacs distributed throughout the body. The air sacs work as bellows, but they do not carry out gas exchange. Oxygen enters the bloodstream through the parabronchi, which are much like the alveoli in human lungs, in that they serve as the tissue for gas exchange. In the parabronchi, blood and air flow past each other in tiny passages. Because air flows in one direction through the lungs, and blood in the lung capillaries flows in the opposite direction, birds can make use of all the air they inhale, much like fish can with their gills and in contrast with mammals, which cannot.

THE AIR SACS

Lung   Abdominal air sac

Cervical air sac

Interclavicular air sac

Posterior thoracic air sac

Anterior thoracic air sac

Section of the lung
The reticulum formed by the parabronchi facilitates the exchange of gases with the blood.

20%
THE PERCENTAGE OF THE BODY'S VOLUME TAKEN UP BY LUNGS AND AIR SACS

1 **INHALATION**
The air sacs fill up with air.

◀ AIR

Anterior air sacs with inhaled air

Posterior air sacs with new air

LUNG

2 **EXHALATION**
The lungs fill up with air.

▶ AIR

Empty anterior air sacs

Empty posterior air sacs

LUNG

Right Jugular

Right Carotid

Right Superior Vena Cava

Right Atrium

Right Ventricle

Left Superior Vena Cava

Left Atrium

Left Ventricle

Aorta

A Highly Complex Heart

Similar to that of reptiles, but having a heart with four chambers instead of three, the circulatory system distributes nutrients and oxygen throughout the body according to the body's needs. The heart's size and rate vary, depending on the bird's weight and activities. In general, bigger birds have smaller and slower hearts. For example, the heart of a seagull on the ground beats 130 times a minute; in flight, it beats 625 times a minute. A hummingbird's heart can beat 700 times a minute.

THE HEART'S ASYMMETRY

The left side of the heart is more developed, because it pumps blood to the whole body. The right side pumps blood to only the lungs.

1 **THE BLOOD**
enters through the right and left arteries.

2 **RELAXED VENTRICLES**
They open the atrioventricular valves.

3 **CONTRACTED VENTRICLES**
The blood enters the bloodstream.

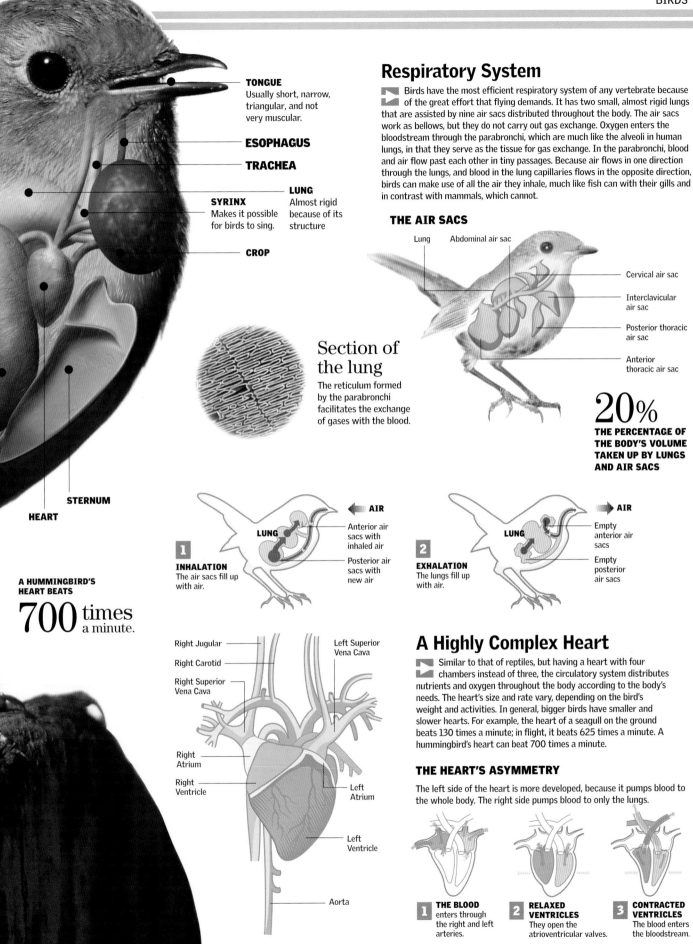

The Senses

In birds, the sense organs are concentrated on the head, except for the sense of touch, which is found all over the body. Birds have the largest eyes with respect to the size of their bodies. This enables them to see distant objects with considerable precision. Their field of vision is very broad, over 300 degrees, but in general they have little binocular vision. The ear—a simple orifice, but very refined in nocturnal hunters—helps them notice sounds inaudible to humans, which facilitates the detection of prey while flying. The senses of touch and smell, on the other hand, are important only to some birds, and the sense of taste is almost nonexistent.

The Ear

Birds' ears are simpler than those of mammals: a bird's ear has no outer portion, and in some cases it is covered with rigid feathers. A notable part of the ear is the columella—a bone that birds share with reptiles. The ear is nonetheless well developed, and birds have very acute hearing; whereas human beings can detect just one note, birds can detect many. The ear is essential to a bird's balance, a key factor in flying. It is also believed that in certain species the ear works as a barometer, indicating altitude.

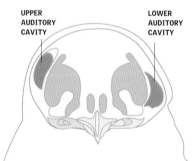

UPPER AUDITORY CAVITY

LOWER AUDITORY CAVITY

LOCATION OF THE EARS

Located at different heights on the head, the ears cause the sense of hearing to occur with a slight delay. In nocturnal hunters, such as owls, this asymmetry allows for the triangulation of sounds and the tracking of prey with a minimal margin of error.

Touch, Taste, and Smell

The sense of touch is well developed in the bill and tongue of many birds, especially in those birds that use them to find food, such as shorebirds and woodpeckers. Usually the tongue is narrow, with few taste buds, but they are sufficient to distinguish among salty, sweet, bitter, and acidic tastes. The sense of smell is not very developed: although the cavity is broad, the olfactory epithelium is reduced. In some birds, such as kiwis and scavengers (condors, for example), the olfactory epithelium is more developed.

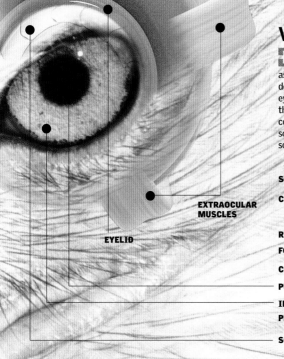

EXTRAOCULAR
MUSCLES

EYELID

Vision

▶ Vision is the most developed sense in birds because some flight maneuvers, as well as the recognition of food from afar, depend on it. Birds have relatively large eyes. In most cases, they are wider than they are deep because the lens and the cornea—which is supported by a series of sclerotic bony plates—project beyond the eye socket. In hunting birds, the eyes are almost tubular. The muscles around the eye change its shape, alter the lens, and create greater visual acuity: birds typically have a 20-fold magnification (and sometimes, as in the case of some diving birds, a 60-fold magnification), in comparison with humans. Their sensitivity to light is also remarkable, with some species being able to recognize light spectra invisible to the human eye.

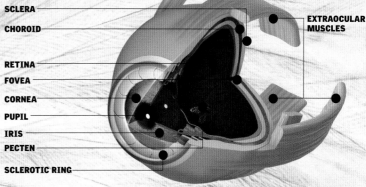

SCLERA
CHOROID
RETINA
FOVEA
CORNEA
PUPIL
IRIS
PECTEN
SCLEROTIC RING

EXTRAOCULAR
MUSCLES

Field of Vision

The eyes—when located on the sides of the head, as is the case with most birds—create a broad field of vision: more than 300 degrees. Each eye covers different areas, focusing on the same object only when looking ahead through a narrow binocular field of vision.

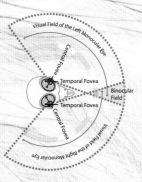

Visual Field of the Left Monocular Eye
Central Fovea
Temporal Fovea
Binocular Field
Temporal Fovea
Central Fovea
Visual Field of the Right Monocular Eye

THE HUMAN FIELD OF VISION

The eyes, located at the front, move together, covering the same area. Because human beings cannot move their eyes independently from each other, they have only binocular vision.

Visual Field of the Left Monocular Eye
Binocular Field
Visual Field of the Right Monocular Eye

COMPARISON OF BINOCULAR FIELDS OF VISION

Binocular vision is essential for measuring distances without making mistakes. The brain processes the images that each eye generates separately, as if they were a single image. The small differences between the two images allow the brain to create a third one in depth, or in three dimensions. Hunting birds, for which the correct perception of distance is a life-and-death matter, tend to have eyes located toward the front, with a wide field of binocular vision. In contrast, birds with lateral eyes calculate distance by moving their heads, but they record a larger total field of vision to avoid becoming prey. Owls are the birds with the greatest binocular vision—up to 70 degrees.

HUNTING BIRDS' FIELD OF VISION

Frontal eyes reduce the total field of vision but allow for a wide field of binocular vision.

A   B

BINOCULAR
FIELD OF
VISION

MONOCULAR FIELD
OF VISION

NONHUNTING BIRDS' FIELD OF VISION

The lateral eyes open the field of vision to as much as 360 degrees but reduce the binocular field.

A   B

MONOCULAR
FIELD OF VISION

BINOCULAR
FIELD OF
VISION

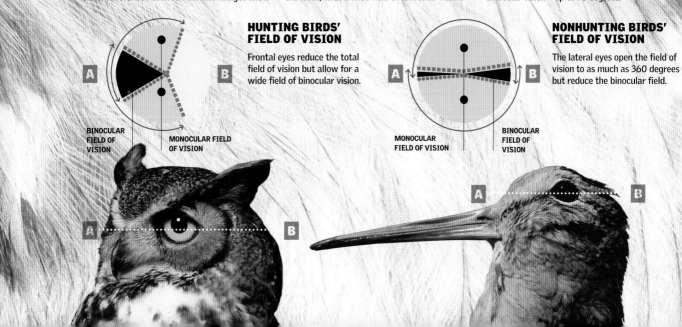

A   B

A   B

Feathers

Feathers are the feature that distinguishes birds from all other animals. They make birds strikingly colorful, protect them against cold and intense heat, enable them to move easily through the air and water, and hide them from enemies. Feathers are also one of the reasons why human beings have domesticated, caught, and hunted birds. A bird's set of feathers is called its plumage, and its color is essential for reproductive success.

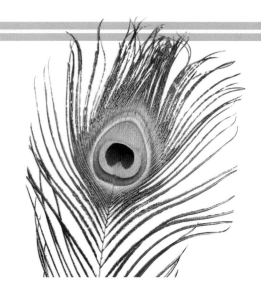

Structure

The structure of feathers has two parts: a shaft and a blade. The shaft is called the rachis, and the part connected to the bird's skin is called the calamus. The movement of a feather is generated in the rachis. The blade is composed of barbs that branch into barbules. The feather's blade, in which the barbules have a series of barbicels, or hooklets, at the tip, is called a vane. The interlocking hooklets in the vane create a network that adds rigidity and resistance to the feather. It also defines the characteristic aerodynamic shape of feathers and helps make the feather waterproof. When feathers wear out, birds have the ability to replace them with new ones.

1 A swelling, or papilla, develops in the bird's skin.

2 In the papilla, special skin cells form a follicle.

3 A tube that will extend from its base and become a feather grows in the follicle.

EDGE
The edge presents an excellent aerodynamic profile for flying.

RACHIS
A feather's main shaft, similar to a hollow rod

HOLLOW INTERIOR

INNER PULP OF THE SHAFT

INFERIOR UMBILICUS
The orifice at the base of the calamus, into which the dermic papilla penetrates. New feathers receive nourishment through it.

CALAMUS
It provides the necessary nutrients for feathers to grow. Nerve endings that stimulate the feather's movement are found at its base. This allows the bird to detect changes in its surroundings.

SUPERIOR UMBILICUS
It contains some loose barbs. Some feathers have a secondary rachis, the hyporachis.

BARBS
are slim, straight ramifications that grow perpendicular to the rachis.

Types of Feathers

There are three main types of feathers, classified according to placement: those closest to the body are down, or underlying feathers; those at the top are contour feathers; and those on the wings and tail are flight feathers, which are often referred to as remiges (on the wings) and rectrices (on the tail).

DOWN
These light and silky feathers protect the bird against the cold. They have a short rachis, or none at all. Their barbs are long, and their barbules lack hooklets. In general, down is the first type of feather that birds develop when they hatch.

CONTOUR
Also called covert feathers, they are short and rounded. They are more rigid than down feathers. Because they cover the body, wings, and tail, they give birds their shape as they fly.

What Is Keratin?

Keratin is a protein that forms part of the outermost layer of a bird's skin, just as it does in other vertebrate animal groups. Keratin is the main component of feathers, hair, and scales. Its distinct resistance helps keep the hooklets woven together in the vane. This allows birds' feathers to maintain their shape in spite of the pressure exerted by the air during flight.

BARBS

BARBULES

HOOKLETS, OR BARBICELS

VANE, OR BLADE
Its outer portion contains a great number of barbicels.

25,000

THE NUMBER OF FEATHERS THAT LARGE BIRDS, SUCH AS SWANS, CAN HAVE.
In contrast, the number of feathers small birds, such as songbirds, can have varies between 2,000 and 4,000.

TRAILING EDGE NOTCH
The turbulence during flight is reduced by this notch, found near the tip of the wing.

PREENING THE PLUMAGE

Birds need to preen their feathers with their bills not only to keep them clean and free of parasites but also to keep them lubricated, which helps birds resist inclement weather. Birds touch their uropygial, or preen, glands with their bills. Then they distribute the oil and wax this gland produces all over their plumage. This task is a matter of survival.

SELF-CLEANING WITH ANTS

Some birds, such as certain tanagers, catch ants with their bills and grind them. They then oil their feathers with the ground-up ants. It is believed that the acid juices from the squashed ants work as a repellent against lice and other external parasites.

DUST BATH

Birds such as pheasants, partridges, ostriches, pigeons, and sparrows perform dust baths to control the amount of grease on their feathers.

PTERYLAE AND APTERIA

At first glance, a bird's body is covered with feathers. However, feathers do not grow all over the body but rather in particular areas called pterylae. This is where the papillae, which create new feathers, are found. The shape and placement of pterylae vary according to species. Pterylae are surrounded by naked areas, called apteria, in which feathers do not grow. Penguins are the only birds whose bodies are completely covered with feathers. This characteristic makes it possible for them to live in cold regions.

IMPERIAL HERON
Powder down keeps its plumage waterproof.

SPECIAL FEATHERS

Vibrissae are special feathers formed by only one filament. Sometimes they have loose barbs at the base that perform a tactile function. They are located at the base of bills or nostrils or around the eyes. They are very thin and are usually blended with contour feathers.

Vibrissae

Filoplumes

POWDER DOWN

This special type of feather can be found on some aquatic birds. They grow constantly and break off at the tip into small waxy scales. This "powder" is preened into the plumage to provide protection.

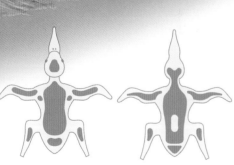

Wings to Fly

Wings are highly modified arms that, through their unique structure and shape, enable most birds to fly. There are many types of wings; they vary by species. For instance, penguins, which are flightless, use their wings for the specialized task of swimming. Among all wings that have existed in the animal kingdom, those of birds are the best for flying. Their wings are light and durable, and in some cases their shape and effectiveness can be modified during flight. To understand the relationship between wings and a bird's weight, the concept of wing loading, which helps explain the type of flight for each species, is useful.

Wings in the Animal Kingdom

Wings have always been modified arms, from the first models on pterosaurs to those on modern birds. Wings have evolved, beginning with the adaptation of bones. Nonavian wings have a membranous surface composed of flexible skin. They extend from the bones of the hand and body usually down to the legs, depending on the species. Avian wings, on the other hand, are based on a very different principle: the arm and hand form a complex of skin, bone, and muscle, with a wing surface consisting of feathers. Furthermore, the avian wing allows for important changes in form, depending on the bird's adaptation to the environment.

HAND BONES

SKIN

PTERODACTYLS
still had talons, and only one finger extended their wings.

HAND BONES

SKIN WITH HAIR

BATS
Four fingers extend the membrane, and the thumb remained as a talon.

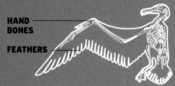

HAND BONES

FEATHERS

BIRDS
The fused fingers form the tip of the wing where the rectrices, or primary feathers, are attached.

Types of Wings

According to the environment in which they live and the type of flight they perform, birds have different wing shapes that allow them to save energy and to perform efficiently during flight. The wing shape also depends on the bird's size. Consequently, the number of primary and secondary feathers changes, depending on the needs of a given species.

THE EXTERNAL PRIMARY FEATHERS ARE LONGER.

THE OUTERMOST PRIMARY FEATHERS ARE SHORTER THAN THE CENTRAL ONES.

THEY ARE WIDE AT THE BASE, WITH SEPARATE FEATHER TIPS.

THERE ARE MANY SECONDARY FEATHERS.

SHORT FEATHERS ARE LOCATED ALL OVER THE WING.

  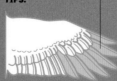 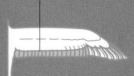 

FAST WING
Remiges are large and tight to allow for flapping; the surface is reduced to prevent excessive friction.

ELLIPTICAL WINGS
Functional for mixed flights, they are very maneuverable. Many birds have them.

WINGS FOR SOARING ABOVE LAND
Wide, they are used to fly at low speeds. The separate remiges prevent turbulence when gliding.

WINGS FOR SOARING ABOVE THE OCEAN
Their great length and small width make them ideal for gliding against the wind, as flying requires.

WINGS FOR SWIMMING
In adapting to swimming, the feathers of penguins became short, and they serve primarily as insulation.

Wing Size and Loading

The wingspan is the distance between the tips of the wings. Together with width, it determines the surface area, which is an essential measurement for bird flight. Not just any wing can support any bird. There is a close relationship between the animal's size (measured by weight) and the surface area of its wings. This relationship is called wing loading, and it is crucial in understanding the flight of certain species. Albatrosses, with large wings, have low wing loading, which makes them great gliders, whereas hummingbirds have to flap their small wings intensely to support their own weight. The smaller the wing loading, the more a bird can glide; the bigger, the faster a bird can fly.

11½ FEET
(3.5 M)

ARGENTAVIS MAGNIFICENS (EXTINCT)

24 FEET
(7.3 M)

WANDERING ALBATROSS

5 FEET
(1.5 M)

LARGER FINGER

SMALLER FINGER

CARPOMETACARPUS

ALULAR DIGIT
Controls the alula, a feathered projection on the front edge of the wing.

ULNA

RADIUS

HUMERUS

CORACOID

STERNUM
OR KEEL

PRIMARIES
They are in charge of propulsion; they are also called remiges.

PRIMARY COVERTS
They cover the remiges and, with the alula, change the wing shape at will.

MEDIAN WING COVERTS
They change the wing's lift when they rise slightly.

SECONDARIES
Their number varies greatly depending on the species. They complete the surface.

GREATER WING COVERTS
They create more surface area and cover the intersection point of the tertiaries.

TERTIARIES
Together with the secondaries, they create the wing's surface.

LOOSE FEATHERS
Sometimes barbicels are missing, and feathers on the wing come apart, creating a loose and ruffled appearance.

PRIMARY FEATHERS
Flying birds have from 9 to 12 primary feathers. Running birds may have up to 16.

Flightless Wings

Among these, penguins' wings are an extreme case of adaptation: designed for rowing underwater, they work as fins. On running birds, wings' first and foremost function is to provide balance as the bird runs. These wings are also related to courtship, as birds show off their ornamental feathers during mating season by opening their wings or flapping them. Wings are also very efficient at controlling temperature, because birds use them as fans to ventilate their bodies.

FUNCTION
The wings of ostriches carry out the functions of balancing, temperature regulation, and courtship.

Types of Tail

Over the course of evolution, birds' tail vertebrae fused into a pygostyle, and in their place feathers of different sizes and colors emerged. These feathers have multiple uses: they can control aerial maneuvers during flight, work as brakes during landing, and make noise. Males also use them during courtship to dazzle and win over females. Usually, the tail is formed by rectrices that vary in number, length, and rigidity, depending on the species.

The Key to How It Works

The tail can perform a variety of functions because of the movement and shape of the feathers. The powerful muscles in the pygostyle prepare the plumage for courtship displays and for flight, provide balance in walking and alighting on trees, and work as rudders for swimming.

OPEN

CLOSED

OPEN

LANDING I
The plumage spreads out, and the main axis of the body is positioned parallel to the ground.

LANDING II
The body leans backward, and the tail closes. The legs prepare to grab the branch.

LANDING III
The spread-out tail feathers, together with an intense flapping of the wings, make it possible for the bird to slow down and prepare its body to land.

Courtship Display

The tail feathers of the female black grouse are straight, whereas those of the male have a half-moon shape. They usually keep the feathers closed and near the ground, but during the courtship displays, they spread them out and show them off completely. To finish the show, the male runs back and forth in front of the female.

OPEN

CLOSED

RECTRICES
Tail feathers can wear out and fray because of friction during flight or by brushing against vegetation.

UNDERTAIL COVERTS
Feathers that cover the lower part of rectrices, protecting them against the wear and tear caused by air friction

BLACK GROUSE
Lyrurus tetrix
The male is recognized by its bluish black plumage and the red caruncle over its eyes.

Fan of Rectrices

On flying birds, it is light and aerodynamic. On tree-climbing birds, such as woodpeckers, the plumage is rigid, which allows them to use it as a support (pointed tail). The coverts of male peacocks are more developed than their rectrices so that the peacock can show them off.

FORKED TAIL
Found on swallows and frigate birds. The external feathers are very long and look like scissors.

ROUNDED TAIL
Found on some songbirds. The central feathers are only slightly longer than the external ones.

GRADUATED TAIL
Found on trogons and kingfishers. When closed, the tail has a layered shape.

MARGINATED TAIL
Found on blue jays. The central feathers are only slightly shorter than the external ones.

SQUARE TAIL
Found on quails. The tail is short, with even feathers.

Gliding

Involves using air currents to fly and save energy when traveling long distances. There are two types of gliders, terrestrial birds and marine birds, each of which is adapted to different atmospheric phenomena. Terrestrial birds rise on thermals (rising air currents). Marine birds use oceanic surface winds. Once the birds gain altitude, they glide off in straight paths. They slowly lose altitude until encountering another thermal that will lift them. Both terrestrial and marine gliders have wings of considerable size.

TYPES OF GLIDING FEATHERS

TERRESTRIAL GLIDER
A large wing surface allows it to make the most of rising air currents at moderate speed.

MARINE GLIDER
Thin and long wings allow it to make the most of the constant surface winds and offer less resistance to forward movement.

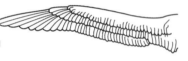

Takeoff

Usually, a powerful jump followed by the vertical flapping of the wings is enough to make a bird take flight. As the wing descends, the tip feathers are stacked on top of each other, forming an airtight surface that helps drive the bird upward. As the bird raises its wings to repeat the movement, the feathers curve and open until the wing reaches its highest point. With a couple of flaps of the wings, the bird is in flight. Bigger birds need a running start on the ground or water in order to take off.

AIR

2 During the downward movement, the primary feathers are closed, which prevents air from passing through.

3 Ascent

FAST AND STRONG FLAPPING

1 During the upward movement in wing flapping, the primary feathers open up, offering less resistance to the air.

INITIAL JUMP

RUN

SECONDARY FEATHERS
There are many of these because of the wing's length.

THE WING LENGTH OF SOME PELICANS MAY REACH 8 FEET (240 CM) FROM TIP TO TIP.

PRIMARY FEATHERS
There are fewer of these, because they only form the tip.

70%
THE ENERGY SAVED BY A SEAGULL WHILE GLIDING

MOVING FORWARD

CONTINUOUS AIR

The tip feathers work as airplane winglets.

WINGLETS

Terrestrial gliders usually have separate primary feathers (toward the tip of the wing) that serve to decrease the noise and tension generated there by the passing of air. Modern airplanes copy their design.

Airplane Winglets are made of one or several pieces.

Marine Birds

Dynamic soaring is performed by birds with long and thin wings, such as the albatross. These wings are designed to take advantage of horizontal air currents, which are responsible for the formation of waves in the ocean. The result is a flight consisting of a series of loops as the bird is lifted upward when it faces the wind and moved forward when it faces away from the wind. This kind of flight can be performed at any time.

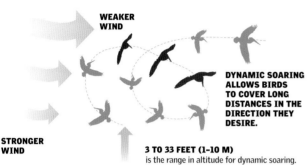

WEAKER WIND

DYNAMIC SOARING ALLOWS BIRDS TO COVER LONG DISTANCES IN THE DIRECTION THEY DESIRE.

STRONGER WIND

3 TO 33 FEET (1-10 M)
is the range in altitude for dynamic soaring.

FLIGHT PATTERNS

Flying in formation is a way for birds in flapping flight to save energy. The leader encounters more resistance as it flies, while the others take advantage of its wake. There are two basic patterns: "L" and "V." The first is used by pelicans, and the second is used by geese.

RELAY
When the leader gets tired, another bird takes its position.

14%
THE PERCENTAGE OF WING FLAPPING THAT GEESE SPARE THEMSELVES BY FLYING IN FORMATION

"L" FORMATION
LEADER
The leader makes the most effort, as it "parts" the air.

THE REST OF THE FORMATION
The other birds use the turbulence produced by the leader's flapping to gain height, following along behind.

"V" FORMATION
The principle is the same, but the birds form two lines that converge at a point. This is the usual formation used by geese, ducks, and herons.

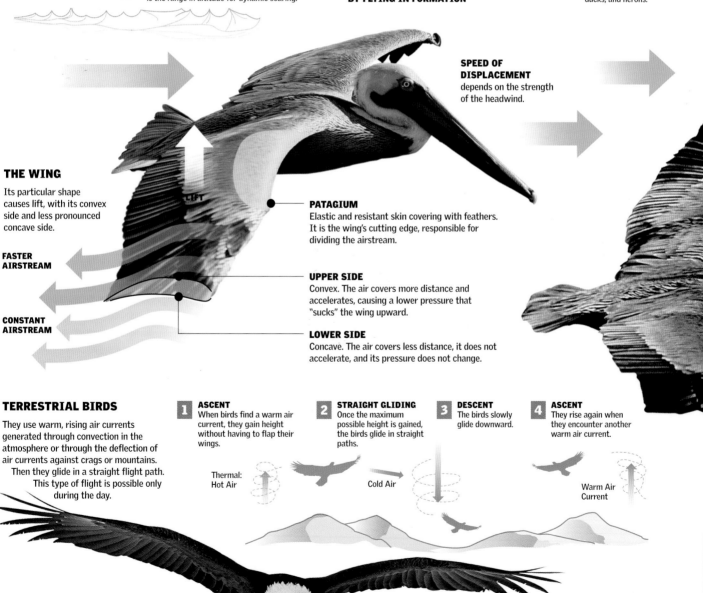

SPEED OF DISPLACEMENT
depends on the strength of the headwind.

THE WING
Its particular shape causes lift, with its convex side and less pronounced concave side.

FASTER AIRSTREAM

CONSTANT AIRSTREAM

PATAGIUM
Elastic and resistant skin covering with feathers. It is the wing's cutting edge, responsible for dividing the airstream.

UPPER SIDE
Convex. The air covers more distance and accelerates, causing a lower pressure that "sucks" the wing upward.

LOWER SIDE
Concave. The air covers less distance, it does not accelerate, and its pressure does not change.

TERRESTRIAL BIRDS

They use warm, rising air currents generated through convection in the atmosphere or through the deflection of air currents against crags or mountains. Then they glide in a straight flight path. This type of flight is possible only during the day.

1 ASCENT
When birds find a warm air current, they gain height without having to flap their wings.

Thermal: Hot Air

2 STRAIGHT GLIDING
Once the maximum possible height is gained, the birds glide in straight paths.

Cold Air

3 DESCENT
The birds slowly glide downward.

4 ASCENT
They rise again when they encounter another warm air current.

Warm Air Current

Flapping Flight

Most flying birds use flapping flight all the time. It consists of moving through the air as if rowing with the wings. With each flap (raising and lowering), the wing both sustains the bird in the air and pushes its body forward. There are different types of flapping flight and different rates of flapping. In general, the larger the bird, the more powerful and less frequent its flapping will be. Because flapping is an activity that consumes much energy, birds have adapted a variety of flight patterns: some, like hummingbirds, always flap their wings, whereas others alternate flapping with short-term gliding. The wing shape also varies according to the bird's needs. Birds that cover long distances have long, narrow wings; those that fly among trees have short, rounded wings.

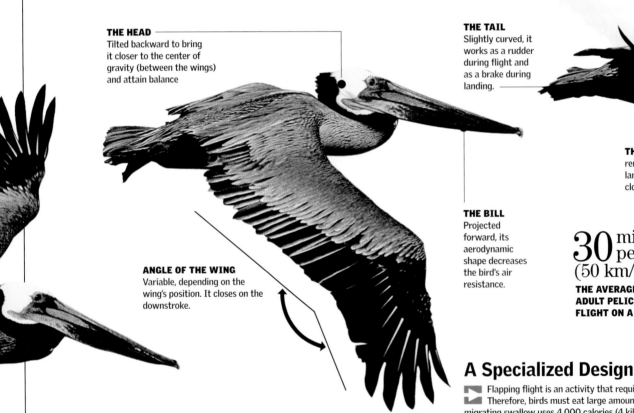

THE HEAD
Tilted backward to bring it closer to the center of gravity (between the wings) and attain balance

THE TAIL
Slightly curved, it works as a rudder during flight and as a brake during landing.

THE LEGS
remain at rest until landing. They stay very close to the body.

THE BILL
Projected forward, its aerodynamic shape decreases the bird's air resistance.

ANGLE OF THE WING
Variable, depending on the wing's position. It closes on the downstroke.

30 miles per hour (50 km/h)
THE AVERAGE SPEED OF AN ADULT PELICAN DURING FLIGHT ON A WINDLESS DAY

A Specialized Design

Flapping flight is an activity that requires much effort. Therefore, birds must eat large amounts of food. A migrating swallow uses 4,000 calories (4 kilocalories) per 1.6 miles (2.5 km) of flight, whereas a small mammal needs only about 25 calories (0.025 kilocalorie) to travel the same distance.

WAVELIKE FLIGHT PATH

Ideal for high speeds, it consists of flapping the wings to gain height and then folding them in order to descend along the flight's trajectory. Afterward, the bird flaps its wings again, making use of the inertia of its descent to regain height. A variation of this type of flight involves gliding between flaps of the wings.

1 PROPULSION
The bird flaps its wings to ascend.

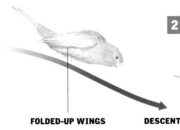

2 REST
The bird keeps the wings near its body to save energy for short intervals.

ASCENT · **FLAPPING WINGS** · **FOLDED-UP WINGS** · **DESCENT**

① Upstroke

As the wings move upward, the remiges separate and form grooves to reduce friction. Support for the bird comes from the patagium, a layer of skin that anchors the feathers and covers the bones.

STRENGTH

To gain height above the ground, the wings flap in big arches in a manner that generally produces a lot of noise.

② Downstroke

As the wings move downward, the remiges are forced together, and the wing moves forward a little for extra support. The wing also bends at the tips to push the bird forward, as if it were rowing.

Muscular strength is distributed to the entire wing, but it increases near the tip.

The downstroke of the wing provides propulsion.

THE CROP
Made of elastic skin. It can hold food during flight.

WING STROKE
The wing acts like an oar as it traps air and pushes the bird forward.

Landing

Requires reducing speed until the bird becomes motionless and settles. The bird faces the wind and spreads out its tail, wings, and alulae (bastard wings, characterized by their stiffness and growth from the first digit), while lifting up its body and extending its legs forward to increase the surface area in contact with the air. In addition, the bird flaps its wings intensely in the direction opposite to its flight. Everything works like an aerodynamic brake. Some birds—such as the albatross, with its long, narrow wings—tend to have problems slowing down. As a result, they are ungainly when landing on the ground, but on the water they are able to ski on their feet until coming to a stop.

WINDMILL FLIGHT: HUMMINGBIRDS

Hummingbirds are able to hover in order to suck the nectar out of flowers. In contrast to other birds, hummingbirds' wings are attached only at the shoulders, which provides greater freedom of wing movement, allowing the hummingbird to hold itself in the air during both the upstroke and the downstroke. The hummingbird has to flap its wings up to 4,800 times per minute during directional flight and for hovering.

The wing has short, sturdy bones; the muscles are very powerful.

Diagram describing the movements of the tip of the wing during flight

THE WINGS FLAP 80 TIMES PER SECOND DURING NORMAL FLYING.

COURTSHIP DISPLAY
Certain hummingbird species can flap their wings up to 200 times per second during courtship.

Great Maneuverability: Hummingbirds are the only birds capable of moving backward.

FLAPPING AGAINST THE WIND

WIND

SPREAD TAIL

SLIDING

The feet spread open before landing to provide more resistance and help the bird to slow down.

The Lives of Birds

The behavior of birds is closely connected to the seasons. To survive, birds must prepare for the arrival of fall and winter and adjust their behavior accordingly. Gliding over the oceans, a wandering albatross, for example, can travel anywhere from 1,800 to 9,300 miles (2,900 to 15,000 km) in a single day in search of food.

PARTRIDGE EGGS
(Lagopus lagopus scoticus)
The female lays eggs at intervals of one to two days, and she is the one who incubates them.

When the time comes to choose a partner, the behavior of males is different from that of females: males employ a variety of tactics to win over females and convince them of their fitness. Some bird couples stay together forever, whereas other birds change partners every year. As for caring for chicks and building nests, in most species both parents participate.

How They Communicate

Sound is an important form of expression in the lives of birds. Birds' sounds can be of two types: calls and songs. The former have a simple acoustic structure, with few notes. They are associated with coordinating the activities of a group, establishing communication between parents and their young, and maintaining contact between birds during migration. Songs, on the other hand, are more complex in rhythm and modulation. They are controlled by the sex hormones, primarily the male hormones. For this reason, males produce the most varied melodies. Songs are linked to sexual behavior and territorial defense. In general, birds either inherit or learn them.

1 The Song and the Brain

Birds have a brain that is well developed for this function. Testosterone acts on the upper vocal center of the brain, which is in charge of memorizing, identifying, and transmitting the orders for the execution of the song.

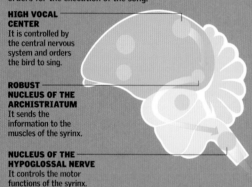

HIGH VOCAL CENTER
It is controlled by the central nervous system and orders the bird to sing.

ROBUST NUCLEUS OF THE ARCHISTRIATUM
It sends the information to the muscles of the syrinx.

NUCLEUS OF THE HYPOGLOSSAL NERVE
It controls the motor functions of the syrinx.

2 Expulsion of Air to the Bronchi

The air stored in the air sacs and lungs is expelled. As it passes through the syrinx (located between the bronchi and the trachea), it vibrates the tympaniform membranes. These membranes are the equivalent of vocal cords in humans.

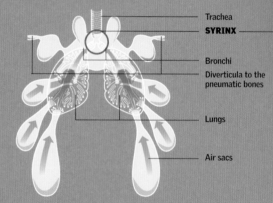

Trachea
SYRINX
Bronchi
Diverticula to the pneumatic bones
Lungs
Air sacs

3 The Production of Sound in the Syrinx

The participation of both the sternotracheal muscles and five to seven pairs of small internal muscles is needed for producing sounds. These muscles control the elongation and contraction of the syrinx, which varies the pitch of the sound. The air sac is also important because it adds external pressure, which causes the tympaniform membranes to tighten. The esophagus works like a resonating box, amplifying the sound. The articulation of the sounds occurs in the buccopharyngeal cavity. There are two types of articulation: guttural and lingual.

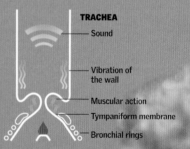

SIMPLE SYRINX
The tympaniform membranes are located above the place where the bronchi divide. They are moved by a pair of external muscles.

TRACHEA
Sound
Vibration of the wall
Muscular action
Tympaniform membrane
Bronchial rings

SONG-PRODUCING SYRINX

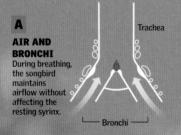

A
AIR AND BRONCHI
During breathing, the songbird maintains airflow without affecting the resting syrinx.

Trachea
Bronchi

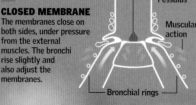

B
CLOSED MEMBRANE
The membranes close on both sides, under pressure from the external muscles. The bronchi rise slightly and also adjust the membranes.

Pessulus
Muscular action
Bronchial rings

C
THE SOUND
The membranes vibrate with the air current and propagate the sound through the trachea until it reaches the bird's bill.

Tympaniform membrane

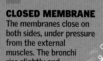

Territoriality and Range

One of the most studied functions of birds' songs is territorial demarcation. When a bird occupies a territory, it sings to announce its claim to competitors, as the pipit shown to the left is doing. When birds must share territory, as in a colony, they develop dialects (variations of sounds produced by the species). When a bird born and raised in one location moves, it must learn the dialect of the new location in order to be accepted and participate in the community. There are also mechanical sounds produced by wing strokes, legs, and bills. In a display of territorial defense, the eared nightjar combines singing with beating its wings.

4,000 bird species

SHARE WITH HUMAN BEINGS AND WHALES THE NEED FOR "SOMEBODY" TO TEACH THEM TO VOCALIZE (SONGBIRDS, HUMMINGBIRDS, AND PARROTS ARE EXAMPLES).

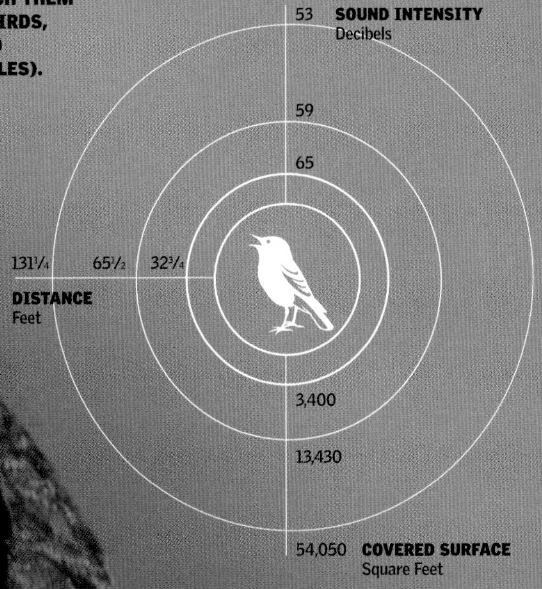

SOUND INTENSITY
Decibels

53
59
65

DISTANCE
Feet

131¼ 65½ 32¾

3,400
13,430

54,050 **COVERED SURFACE**
Square Feet

INTENSITY

can vary widely from bird to bird. The larger the territory, the greater its reach. Its frequencies can change as well: the lower the frequency, the greater the coverage.

Strengthening Ties

Some songbirds develop very complex singing rituals. The duet is perhaps the most striking because it requires both a shared repertoire and good coordination between both birds. In general, the male initiates the song with a repeated introduction; the female then alternates with different phrasing. The phrasing allows for more or less cyclical variations that make it unique. It is believed that this strengthens ties between the pair (as demarcation of territory does) and serves as a stimulus for cooperative behaviors, such as nest building, in which both the male and female may participate.

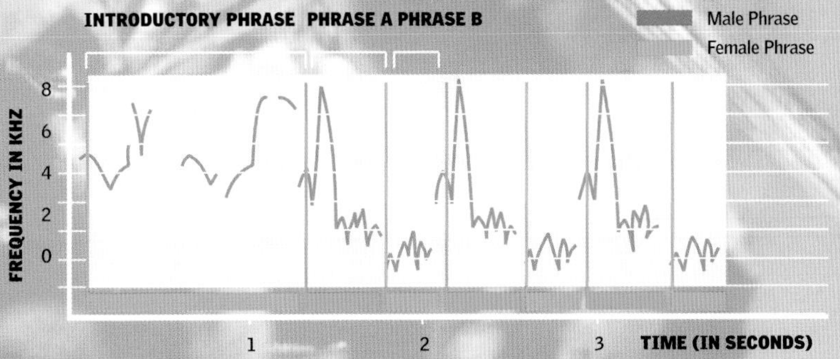

INTRODUCTORY PHRASE PHRASE A PHRASE B

Male Phrase
Female Phrase

FREQUENCY IN KHZ
8 6 4 2 0

1 2 3 TIME (IN SECONDS)

Nuptial Parade

Finding a mate is not easy for any species. For birds, the exhibition of plumage with bright colors, the presentation of offerings and gifts, and the performance of dances and highly elaborate flight patterns are some of the particular behaviors seen during this period. They are known as nuptial or courtship displays. The male resorts to all these strategic gestures to attract the female's attention and prevent her from paying attention to other males. Some of these rituals are extremely complicated; others are very tender and delicate.

A

When its sexual arousal peaks, the male northern harrier (*Circus cyaneus*) flies in a wavelike pattern to attract the female.

B

During courtship, the male northern harrier pretends to attack the female.

AERIAL EXHIBITIONS

Certain birds, such as goshawks or male northern harriers, court the female in flight. They ascend in the air in broad circles, only to let themselves fall in daring, sharp dives.

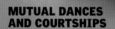

MUTUAL DANCES AND COURTSHIPS

Great crested grebes (*Podiceps cristatus*) perform incredible aquatic dances. They bow to each other, dive, and run through the water side by side.

Special Courtship

Avian courtship is a phenomenon that, depending on the species, can take the form of various rituals. Lek rituals are one of the most intriguing forms of courtship. The males gather in a small area, called an arena, where they perform their courtship displays for the females. The females form a circle around the arena and end up mating with the male that has the most striking secondary sexual characteristics. Lek is a system controlled by the dominant male, who ends up mating with most of the females (polygyny). The less experienced males will mate with only a few, or even none, of the females. For some species, lek rituals can be very intricate. At least 85 species perform this special type of courtship ritual, among them manakins, pheasants, cotingas, and hummingbirds. Manakins, for example, stand in line and wait their turn to perform.

DISPLAYING PHYSICAL ATTRIBUTES

To find a partner, birds such as the snowy egret resort to a series of very elaborate signals, such as songs, poses, dances, flight patterns, noisemaking, and displays of their ornamental feathers.

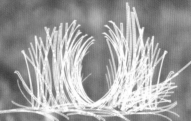

BUILDING BOWERS

Australian bowerbirds build a structure called a bower, which they decorate with pieces of paper and fabric that inevitably attract the female.

GIFTS

Another courtship strategy is the presentation of gifts. Male eagles give females prey, and European bee-eaters offer insects. These offerings are called courtship food.

TIMING

The courtship display is directly related to reproduction cycles. It takes place before copulation, although it can continue to occur thereafter.

PRENUPTIAL
Prenuptial courtship starts with territorial establishment and the search for a partner, which can take place simultaneously.

POSTNUPTIAL
With this display, the great crested grebe ensures the continuity of the pairing even after the eggs are laid.

6 feet
(1.8 m)

IS THE SIZE OF THE TAIL OF THE PEACOCK WHEN IT UNFURLS ITS MORE THAN 200 SHINING FEATHERS AND FORMS A FAN TO ATTRACT THE FEMALE.

GRAY CROWNED CRANE
Balearica regulorum
Two cranes perform a courtship dance consisting of a series of impressive leaps.

EMPEROR PENGUIN
Aptenodytes forsteri
A monogamous species. Each recognizes its partner by its voice, and couples will spend their entire lives together.

Monogamy or Polygamy

Monogamy is the most common mating system, in which two birds, one of each sex, participate, leading to the formation of a couple. This couple can endure for a single reproductive season or for life. Polygamy is an alternative pattern, but it is not very common. Polygamy is divided into two classes: polygyny, in which the male mates with several females, and polyandry, in which the female mates with several males (and may even be able to keep them all together in a harem). In either case, one partner has the sole responsibility of caring for the eggs and chicks. There is also an exceptional case within polygamy: promiscuity. In this arrangement, a couple is not formed, and the relationship is limited to copulation.

Home Sweet Home

Most birds lay their eggs in a nest, where they are incubated by the body heat of an adult sitting on them. To build a nest, the couple normally uses mud mixed with saliva, small stones, branches, and feathers. When the nest is in a visible location, the bird covers it with lichens or loose twigs to hide it from predators. Nest shapes vary according to the bird group: they can take the form of a bowl, a hole in a tree (woodpeckers), or an excavated burrow on a slope of sand or soil. There are even birds that use nests built by other species.

SOME VARIATIONS

WOVEN NESTS
Weaver birds intertwine grass blades until they form a structure. The entrances are underneath.

BURROWS
Parrots and kingfishers dig their nests in sandy river banks.

SEWN NESTS
The tailorbird sews together two large leaves with grass blades. The nest is inside the leaves.

PLATFORM NESTS
The sparrow hawk gathers a large number of branches and assembles a high, solid base for its eggs.

Types and Locations

Nests are classified according to their shape, material, and location. They vary depending on the amount of warmth the species needs, as well as in terms of the level of protection they offer. The greater the pressure from predators, the higher or better hidden a nest must be. Good examples are isolated nests resembling high platforms; nests in deep depressions in the soil or hidden in tree trunks, which are very safe and provide good insulation; and nests made of clay, which are very hard. The most typical nests resemble a cup and are found at various locations, most often between two or three high, remote branches.

How the Nest Is Built

A cup-shaped nest is built at a fork between two or three branches. The bird arranges twigs, grass blades, and small sticks, as if building a platform. The bird then interlaces some of these materials with the tree to give the nest solidity. It then interweaves the materials in a circular pattern. As the nest takes shape, lighter, more adhesive materials—such as mud, spiderwebs, caterpillar silk, and certain plant fibers—are used. Although the outside is rough, the inside is lined with feathers for softness and warmth. In general, if both the male and female participate in the construction, a few hundred trips are enough to complete the nest. In some species, such as weavers, males have to display a nest during courtship; in others (African black eagles, for instance), the same nest is used every year.

The Little Masked Weaver
builds solid nests, weaving together leaves and grass blades. Sometimes the male builds several nests before the female chooses him.

WALLS
These are the most important component in a nest's construction, and they give it its characteristic shape. The materials used vary, depending on the habitat.

BASE
This is the nest's foundation, and it is the first section built by the bird. It is solid and is usually built with bulkier materials.

LINING
It is composed of fibers, hairs, feathers, and down. It insulates the eggs from the cold and helps with incubation.

1 BASE
The bird gathers branches and small sticks on a fork to lay the foundations of the nest, setting up a platformlike structure. The bird then ties it to the tree, making it sturdy.

2 SHAPE
The bird intertwines grass blades, small sticks, and hairs; fastens them; and gives them a circular shape. As construction advances, the bird uses adhesive materials, such as spiderwebs.

3 END
The finish includes moss and feathers to smooth out the interior and provide insulation against the wind and cold, which makes the nest more suitable for incubation.

Structure

The cup shape is important for preventing eggs from rolling out and falling. Besides making the construction of the nest easier, the use of different materials helps make the nest sturdier, because the smaller and more flexible pieces firm up the base, walls, and lining of the structure. Different materials also provide more efficient insulation, keeping warmth in and cold out during the incubation and raising of chicks. As an added reinforcement, birds usually make the side facing the prevailing winds thicker and the side facing the sun thinner. Thus, the whole nest is an incubator that conserves energy. Finally, an exterior finish camouflages the nest among branches, hiding it from predators.

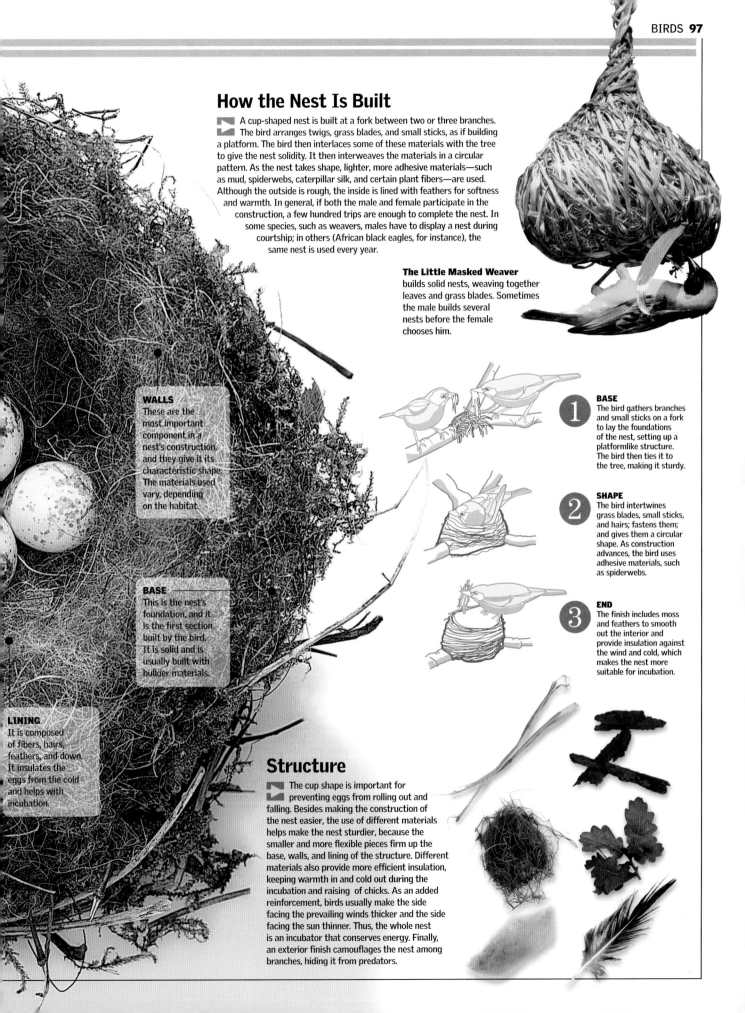

First, the Egg

Birds may have inherited their reproductive method from their predecessors, the theropod reptiles. In general, they lay as many eggs as they can care for until the chicks become independent. Highly adapted to the environment, the eggs of the same species have varying shapes and colors. These variations help keep them safe from predators. They also vary greatly in size: the egg of an ostrich is 2,000 times bigger than that of a hummingbird.

How It Forms

Birds have only one functional ovary, the left one, which grows dramatically during the mating season. The ovule can descend and form what are known as unfertilized eggs (the type used in cooking). If the egg is fertilized, embryonic development begins. The ovule, fertilized or not, descends to the cloaca in a few hours or days. The eggshell begins to be formed at the isthmus, through the secretion of calcium. At first soft, the shell hardens when it comes in contact with the air.

3 Most birds' organs are formed in the first hours of incubation.

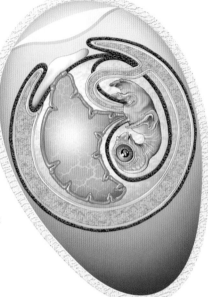

2 As it feeds to grow, the embryo produces wastes that are kept in a special sac.

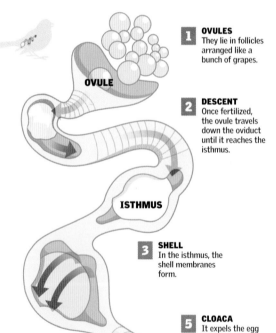

1 OVULES
They lie in follicles arranged like a bunch of grapes.

OVULE

2 DESCENT
Once fertilized, the ovule travels down the oviduct until it reaches the isthmus.

ISTHMUS

3 SHELL
In the isthmus, the shell membranes form.

4 UTERUS
The egg becomes pigmented, and the shell hardens.

CLOACA

5 CLOACA
It expels the egg 24 hours later on average (chicken hens).

WASTE SAC

CHORION
protects and contains the embryo and its food.

YOLK

YOLK SAC

ALBUMIN

1 The egg contains an embryo in one side of the yolk. The yolk is held in the middle of the white (albumin) by a protein cord that isolates it from the outside world.

EMBRYO

PROTEIN CORD (CHALAZA)

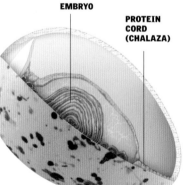

LAYING

A group of eggs laid at one time is called a laying. During the mating season, a sparrow can have several layings. If some eggs are removed, the sparrow can replace them without difficulty.

SHAPE

It depends on the pressure exerted by the oviduct walls. The large end emerges first.

Oval: The most frequent

Conical: Prevents falling

Spherical: Reduces the surface area

COLOR AND TEXTURE

Both texture and color help parents locate the egg.

Light Egg

Dark Egg

Speckled Egg

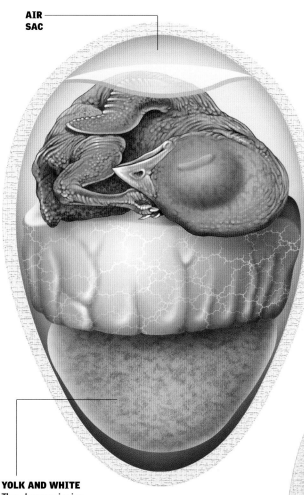

AIR
SAC

YOLK AND WHITE
They decrease in size.

4

The bill and scales of the legs harden toward the end, when the chick is formed and reaches a size similar to that of the egg. At that point, rotation begins so that the chick will be positioned to break its shell.

SIZE
There is no exact proportion between the size of a bird and its egg.

1 pound
(500 g)
KIWI EGG

2 ounces
(60 g)
CHICKEN HEN'S EGG

5

When the chick is ready to break the shell, it is taking up all the space inside the egg. The chick is cramped with its legs against its chest. This enables it to open the shell with small movements and with the help of a hard point at the tip of its bill (called an egg tooth).

THE SHELL

Formed by a solid layer of calcium carbonate (calcite), it has pores that make it possible for the chick to breathe. Bacteria are kept out by two membranes that cover the egg, one on the inside and the other on the outside.

PORE   MEMBRANE   OUTER
AND INNER
MEMBRANES

OXYGEN

CO_2 AND
WATER
VAPOR

8%
**THE PROPORTION OF
AN EGG TAKEN UP BY
THE EGGSHELL**

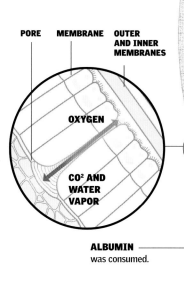

ALBUMIN
was consumed.

YOLK
disappears into the body.

Birth in Detail

When a chick is about to hatch, it starts to make itself heard from inside the egg. This allows it to communicate with its parents. It then starts to peck at the shell with its tiny egg tooth, which is lost after birth. Next, it turns inside the egg and opens a crack with new perforations, at the same time pushing with its neck and legs until it manages to stick out its head. This job demands a lot of effort and can take 30 to 40 minutes or, in the case of kiwis and albatrosses, even three to four days. In most species, newborn chicks are blind and naked, and they can open their bills only to receive food.

INCUBATION

For the embryo to develop, it needs constant temperatures between 99° and 100° F (37–38°C). The parents ensure these temperatures by sitting on the eggs and warming them with their brood patches.

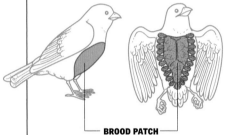

During incubation, some species lose their chest feathers and increase their number of blood vessels in this area. Others pluck out their feathers. Direct contact with the eggs helps keep them warm.

BROOD PATCH

DURATION BY SPECIES
The incubation period varies considerably: between 10 and 80 days, depending on the species.

PIGEON
Females and males incubate. They both develop a brood patch.

18 days

PENGUIN
Both males and females incubate. The emperor male has a special pouch for incubation.

62 days

ALBATROSS
Lacking brood patches, the parents hold the egg between their feet and abdomen.

80 days

Breaking the Shell

This process may take from a few minutes to three or four days, depending on the species. In general, the parents do not intervene or help their young. When the shell is empty, they throw it out of the nest, apparently to avoid attracting the attention of predators. In species whose young hatch with the feathers already developed, hatching is extremely important. It has been observed that the singing of the chicks stimulates the stragglers and delays those that have gotten ahead; it is important that they all be ready to leave the nest together.

35 minutes
IS THE APPROXIMATE TIME IT TAKES A SPARROW TO COME OUT OF THE EGG.

① Crack in the Egg

The chick turns inside until its bill targets the egg's midline. It then punctures the air sac. With a few more tries, it pierces the shell. The chick then breathes for the first time.

ASKING FOR HELP

The chick calls for its parents from inside. The reply encourages it to continue the effort.

SEQUENCE OF PECKING

Between each sequence of pecking, the chick must take long breaks.

Adaptations for Hatching

Getting out of the egg is an intricate operation because the space is tight, and a chick's muscles have little vigor. Birds count on a few adaptations, such as the egg tooth and the hatching muscle, to accomplish the task. The tooth is used for making the first perforation, which allows air into the egg. The muscle exerts the necessary strength, while stimulating the chicken's motor functions to intensify the effort. Both the egg tooth and the hatching muscle disappear shortly after the eggshell is broken.

HATCHING MUSCLE
It exerts pressure against the shell and helps to break it.

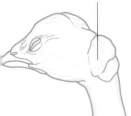

EGG TOOTH
A protuberance on the bill that punctures the egg. Its presence depends on the species.

4

The Chick Is Born

Once outside, the chick, almost featherless, looks for warmth and food from its parents. In the case of some birds that hatch without feathers, not all eggs hatch simultaneously; this benefits the firstborn if food is scarce.

A GREAT EFFORT
Getting out of the shell requires much energy from the chick.

SHELL MEMBRANE

EGGSHELL

3

Getting Out of the Egg

Once the shell is open, the chick pushes itself out with its legs and by crawling on its abdomen. For birds that hatch without feathers, this is more difficult, because they are less developed.

2

The Crack Expands

After making a hole in the shell, the chick opens a crack with successive pecks at other points. Air gets in and dries up the membrane, which makes the task easier.

What Comes Out First?
The head usually comes out first, because the sharp bill helps break the shell. Most birds then get out of the egg by pushing themselves out with their legs. For wading birds and other terrestrial birds, however, the wings usually unfold first.

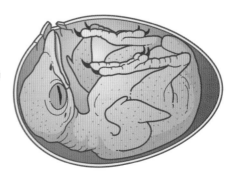

Postnatal Development

Chicks develop at highly variable rates after hatching. Some birds are born with their eyes open and with a thick layer of down feathers. These birds can also feed themselves. That is why they are called precocious, or nidifugous. Ducks, rheas, ostriches, and certain beach birds can walk and swim as soon as they are born. Other species are born naked and develop their feathers later. They need to stay in the nest until they have sufficiently developed, so adult birds must care for them. These birds are called nidicolous. The most helpless chicks are the young of songbirds and hummingbirds, because they need warmth from their parents to grow strong.

Nidifugous Young

Nidifugous young are fully developed at the moment of birth. They can move and even leave the nest, hence their name (which means "fleeing the nest"). This adaptation demands more incubation time because the chick is almost fully developed at birth. This is the case with incubator birds (*Megapodius freycinet*), which begin their independent lives in the outside world as soon as they leave the shell. Ducks follow their parents but find food on their own, whereas chickens follow their parents, and the parents show them where to find food.

EYE
They are born with open eyes.

FEATHERS
The body emerges from the egg covered with damp down. Within three hours, it will become dry and fluffy.

Red-Legged Partridge
Alectoris rufa

MOVEMENT
Within a few hours of hatching, nidifugous birds can run around.

21 days
It is already considered an adult. Its flights are longer. Its diet is composed of 97 percent vegetables; the rest is made up of lichens and insects.

15 days
It starts to perform short flights. It reverses its diet, eating:
66%
seeds and flowers. The rest consists of invertebrates.

GROWTH STAGES

30 hours
The chick keeps warm with the down that covers its body. It can walk and begins to be fed by its parents.

7 to 8 days
Growth quickens, and the first covert feathers appear at the tip of the wing. The bird leaves the nest. Its diet consists of:
66%
invertebrates. The rest consists of seeds and flowers.

24 hours
IS THE MINIMUM AMOUNT OF TIME THE BLACK-HEADED DUCK NEEDS TO BE READY TO FLY.

SIZE COMPARISON

NIDIFUGOUS
The egg is larger, the chick is born more developed, and the incubation period is longer than that of nidicolous birds.

NIDICOLOUS
They lay small eggs, with a brief incubation period, and the young are helpless at birth.

Nidicolous Broods

Most of these chicks are born naked, with closed eyes and with only enough strength to get out of their shells. They stay in the nest. For the first few days, they cannot even regulate their own body temperature; they need their parents in order to stay warm. Within one week, they have a few feathers, but they require constant care and food. They form a numerically important group that includes Passeriformes (songbirds).

FOOD
They need much food to develop. The parents must feed them 24 hours a day.

House Sparrow
Passer domesticus

An adult bird can feed its young up to
400 times a day.

INTERIOR OF THE BILL
Its color is bright to stimulate the parents to regurgitate the food.

—Shining
Areas

Some species have shiny areas that can be seen even in the dark.

EYES
Nidicolous chicks are born blind. They open their eyes a few days after birth.

FEATHERS
The chicks are born either naked or with down feathers in some areas.

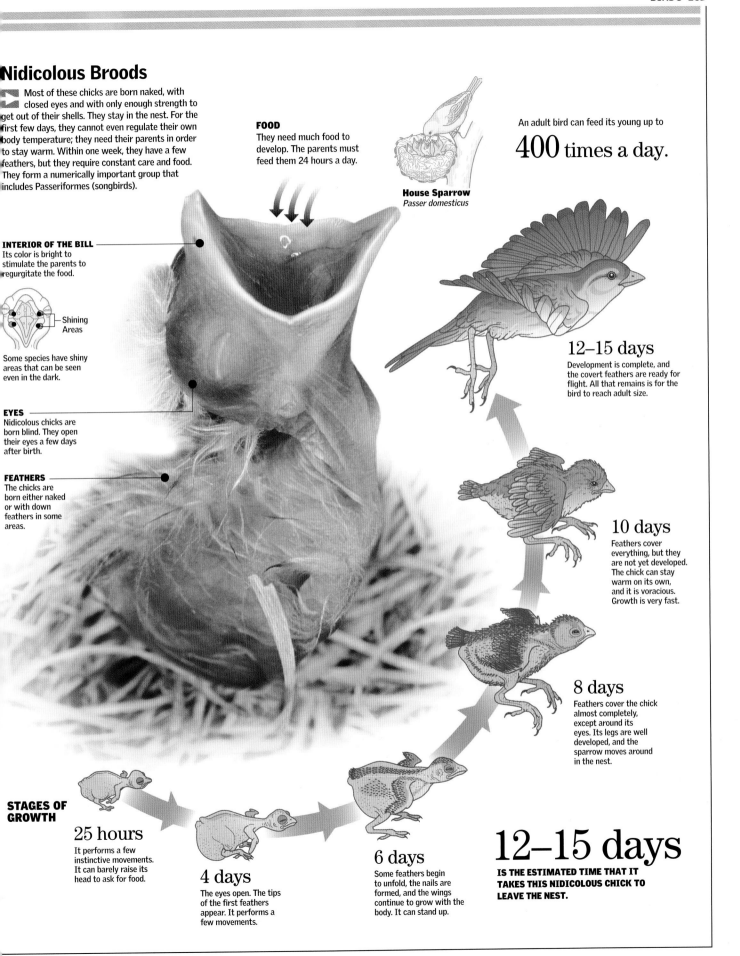

12–15 days
Development is complete, and the covert feathers are ready for flight. All that remains is for the bird to reach adult size.

10 days
Feathers cover everything, but they are not yet developed. The chick can stay warm on its own, and it is voracious. Growth is very fast.

8 days
Feathers cover the chick almost completely, except around its eyes. Its legs are well developed, and the sparrow moves around in the nest.

STAGES OF GROWTH

25 hours
It performs a few instinctive movements. It can barely raise its head to ask for food.

4 days
The eyes open. The tips of the first feathers appear. It performs a few movements.

6 days
Some feathers begin to unfold, the nails are formed, and the wings continue to grow with the body. It can stand up.

12–15 days
IS THE ESTIMATED TIME THAT IT TAKES THIS NIDICOLOUS CHICK TO LEAVE THE NEST.

Diversity and Distribution

The environment in which an organism usually lives is called its habitat. In their habitats, birds find food, the best places to build nests, and escape routes in case of danger. An almost universal pattern of distribution shows that more species live in the tropics than elsewhere. With evolution, birds with a common origin

DUCK
(Anatidae family)
A natural-born fisher,
ducks feed on small snails
and aquatic insect larvae.

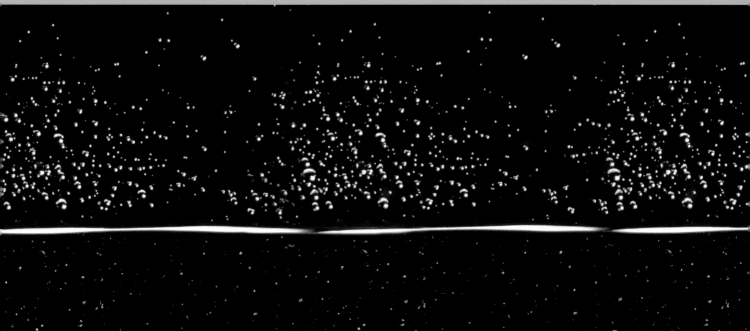

have diversified as they have begun to occupy different environments. This phenomenon is called adaptive radiation. We find ocean birds, which have undergone many changes in order to live near the sea, as well as birds that live in freshwater environments, in forests, and so on. Each type has acquired special physical traits and behaviors as a result of the adaptive process.

Where They Live

With their mobility, birds have conquered all areas of the earth. Despite this characteristic, there are few cosmopolitan species—that is, most birds have specific habitats determined by climate and geographic features. Count de Buffon in the eighteenth century was the first person to notice that living beings are not distributed homogeneously. By analyzing how animals were dispersed on the planet, he realized that different places had different types of fauna. After the work of naturalist Charles Darwin and ornithologist Philip Sclater, it became clear that organisms are situated in specific biogeographic regions.

Puffin
Fratercula artica

NORTH
AMERICA

Atlantic Ocean

Pacific Ocean

CENTRAL
AMERICA

SOUTH
AMERICA

Nearctic
7%

732 Species
62 Families

CHARACTERISTICS

Climatic barrier of cold weather and oceanic isolation

Most migrating species

Many insectivorous and aquatic birds

Affinity with Palearctic

Endemic Avifauna: **loons and puffins**

Oceania
2%

187 Species
15 Families

CHARACTERISTICS

Large area and number of climates

Gliders, divers, and swimmers

Abundance of fish-eating species

Many cosmopolitan species

Endemic Avifauna: **albatrosses, sheathbills, petrels, penguins, and seagulls**

Adaptations According to the Environment

 Birds are found in all habitats of the world, although most live in tropical regions. Their ability to adapt, however, is remarkable. From jungles to deserts, from mountains to coasts, and even on the sea, birds have succeeded in acclimating themselves. They have undergone a highly varied array of changes in form and behavior. Emperor penguins not only nest in Antarctica but they also incubate their eggs between their feet for 62 to 66 days. The male Lichtenstein's sandgrouse has developed a sponge of feathers to bring water to its chicks, and hummingbirds have special wings that enable them to make all kinds of maneuvers.

Neotropic
32%

3,370 Species
86 Families

CHARACTERISTICS

Long-lasting geographic isolation

Many primitive species

Great numbers of frugivores

Endemic Avifauna: **rheas, tinamous, oilbirds, hoatzins, cotingas, and stripe-backed antbirds**

This region undoubtedly has the greatest diversity of birds. The variety in the South American tropics, the most important tropical zone in the world, is one-and-a-half times greater than that of tropical Africa. With more than 1,700 species, Colombia, Brazil, and Peru are the countries with the greatest diversity of avifauna. Even Ecuador, a much smaller country, has more than 1,500 species.

Hoatzin
Opisthocomus hoazin

Biodiversity in the World

 The most diverse regions in terms of bird populations are the tropics because of the favorable conditions of abundant food and warm climate found in them. Temperate regions, however, with their seasons, are destinations for migrating birds from tropical and polar regions. Cold regions, on the other hand, have little diversity but are rich in population density. The rule is that diversity of life-forms happens in places where the environment requires less severe adaptations.

NUMBER OF SPECIES

- up to 200
- 200 – 400
- 400 – 600
- 600 – 800
- 800 – 1000
- 1000 – 1200
- 1200 – 1400
- 1400 – 1600
- 1600 – 1800

Palearctic

937 Species
73 families

CHARACTERISTICS

Climatic barrier of cold weather and oceanic isolation

Low diversity of species

Most are migratory species

Many insectivorous and aquatic birds

Endemic Avifauna: **wood grouse, waxwings, flycatchers, cranes**

9 %

ASIA

Because of similar climatic conditions, many authors merge the Palearctic and Neoarctic regions, calling the combined region Holarctica.

EUROPE

AFRICA

Afrotropic

1,950 Species
73 families

CHARACTERISTICS

Maritime and desert isolation

Great number of Passeriformes

Many flightless birds

Endemic Avifauna: **ostriches, turacos, cuckoos**

19 %

Indian Ocean

Indomalaya

1,700 Species
66 families

CHARACTERISTICS

Affinities with the Afrotropical zone

Tropical birds

Many frugivores

Endemic Avifauna: **ioras, pittas, swifts**

16 %

Pacific Ocean

COUNTRIES WITH THE MOST SPECIES

MORE THAN 1,500
Colombia
Brazil
Peru
Ecuador
Indonesia

MORE THAN 1,000
Bolivia
Venezuela
China
India
Mexico
Democratic Republic of the Congo
Tanzania
Kenya
Argentina

Ostrich
Struthio camelus

Ruby-Throated Hummingbird
Archilochus colubris

OCEANIA

Australasia

1,590 Species
64 families

CHARACTERISTICS

Long isolation

Many flightless and primitive birds

Endemic Avifauna: **emus, kiwis, cockatoos, birds of paradise**

15 %

No Flying Allowed

A few birds have lost their ability to fly. Their main characteristic is wing loss or reduction, although for some a remarkable size may be the cause of their inability to fly. Such birds weigh more than 39 pounds (18 kg). This is the case with runners (ostriches, cassowaries, emus, rheas, kiwis); extremely fast birds that live in remote areas of New Zealand; and swimmers, such as penguins, that have developed extraordinary aquatic abilities.

AFRICAN OSTRICH
A single species inhabits eastern and southern Africa. Adults reach a height of 9 feet (2.75 m) and a weight of 330 pounds (150 kg).

Super Swimmers

Penguins' bodies are covered with three layers of small, overlapping feathers. A penguin has small limbs and a hydrodynamic shape that helps it swim with agility and speed. Dense, waterproof plumage and a layer of fat insulate the bird from the low temperatures of the regions where it lives. Because its bones are rigid and compact, it is able to submerge itself easily. This adaptation distinguishes it from flying birds, whose bones are light and hollow.

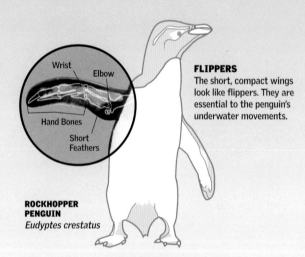

Wrist
Elbow
Hand Bones
Short Feathers

FLIPPERS
The short, compact wings look like flippers. They are essential to the penguin's underwater movements.

ROCKHOPPER PENGUIN
Eudyptes crestatus

SMALL HEAD
LONG NECK
ATROPHIED WINGS
PELVIS
FLAT STERNUM
ROBUST BONE

PENGUIN HEADING TO THE WATER

HUNTING
The wings work like flippers. The foot—with four joined toes pointing backward—and the tail steer the direction of the dive.

BREATHING
When looking for food, penguins need to leave the water and take a breath between plunges.

RELAXING
When resting in the water, they move slowly. They float on the surface with their heads up and balance their bodies with their wings and feet.

RUNNER'S CHEST

The keel-shaped sternum of flying and swimming birds offers a larger surface for attachment of the pectoral muscles. The flat sternum of running birds has a smaller surface and, consequently, less mobility.

KEEL-SHAPED STERNUM

The Ratites

Running birds belong to the group of the ratites (*rata* = raft, an allusion to the flat sternum). The front limbs either are atrophied or have functions unrelated to flying. The hind limbs have very strong muscles as well as sturdy, vigorous bones. Another difference is found in the sternum. It is a flat bone without a keel, which flying and swimming birds possess. Wild ratites can be found only in the Southern Hemisphere. The Tinamidae, native to Central and South America, belong to this group (partridges).

6 feet (1.8 m)

4 feet (1.2 m)

4½ feet (1.4 m)

1¼ feet (0.4 m)

STRUTHIONIFORMES
The ostrich is the only species in this group. It uses its wings for balance when running fast. It has only two toes on each foot. The adult male can weigh up to 330 pounds (150 kg).

RHEIFORMES
Rhea are common in South American countries, such as Argentina. They look like ostriches but are smaller. Their three-toed feet allow them to chase prey. Their long necks and excellent eyesight make them skillful hunters.

CASUARIIFORMES
Agile runners and swimmers. The colors on their necks and heads are distinctive. A bony hoof protects them from vegetation when they run. They have long, sharp talons on their feet.

APTERYGIFORMES
Kiwis. These birds have four toes on each foot, and their feathers look like fur because they do not have barbules. They normally use their keen sense of smell to find insects at night. They lay only one large egg.

ASIA

OCEANIA

NEW ZEALAND

2 feet (0.6 m)

Running and Kicking

Ostriches usually run to escape from predators or to hunt small lizards and rodents. In both cases, because of their strong legs, they are able to reach a speed of 45 miles per hour (72 km/h) and to maintain it for 20 minutes. When running is not enough to protect the bird, kicking is a valid recourse that discourages the attacker. In courtship displays, forceful stamping is also used to win over females.

GREATER DIVERSITY
In many cases, running birds can be found in many parts of the world because of human intervention. The area where flightless birds have diversified the most is Oceania, due to continental isolation.

18
VERTEBRAE
THE NUMBER AN OSTRICH HAS IN ITS NECK

TARSUS METATARSUS

PHALANGEAL CUSHION

PHALANGES

CLAW

TOE

PLANTAR CUSHION

On Two Toes
With just two toes, the contact surface between the foot and the ground is relatively small. This is an advantage when moving on land.

Other Walkers

More than 260 species belong to the order Galliformes, which includes chickens, turkeys, and pheasants. The birds in this group have keels, and they perform abrupt and fast flights, but only in extreme situations. Their feet are suitable for walking, running, and scratching the ground. This group includes the birds that human beings use the most. In general, males are in charge of incubating and raising the young.

1 Taking a run and jump

2 Clumsy and flapping fast

3 Emergency landing

Marine Residents

Of the more than 10,000 bird species inhabiting the earth, only about 300 have managed to adapt to marine life. To survive at sea, they have undergone multiple adaptations. For instance, marine birds have a more efficient excretory system than that of other species, including a specific gland that helps them eliminate excess salt. Most marine birds live on the coasts and have mixed behaviors; others are more aquatic than aerial. A few—such as albatrosses, petrels, and shearwaters—can fly for months at a time, landing only to raise their chicks. They are called pelagic birds.

Adaptations

Marine birds are well prepared to live on water, especially those that fish out at sea. The tips of their bills are hook-shaped, and their feet have webbed membranes between the toes. They also have an admirable ability to float. The saline water is not a problem; these birds can even drink it. In some pelagic birds, sense of smell plays an important role in enabling them to detect the oil of the fish in the water to find schools of fish. They also use their sense of smell to find their nests in colonies.

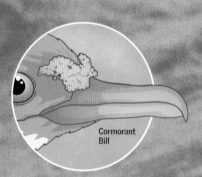

Cormorant Bill

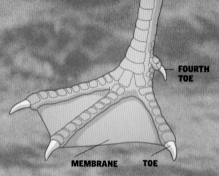

FOURTH TOE

MEMBRANE   **TOE**

TOTIPALMATE FEET
A characteristic of many marine birds. The posterior toe (hallux) is joined to the other toes by a membrane. It creates more surface area, and, therefore, more push, as the bird swims. Birds with this kind of foot walk clumsily.

148 feet (45 m)
IS THE MAXIMUM DEPTH THEY REACH. COMMON LOONS—BIRDS INDIGENOUS TO NORTH AMERICA THAT SPEND THE WINTER AT SEA—HAVE BEEN RECORDED TO REACH THIS DEPTH.

Common Loon
Gavia immer
Almost unable to walk, common loons are excellent swimmers and divers. They nest at inland lakes during the summer.

VARIOUS MARINE BIRDS

HOOK-SHAPED TIP
Prevents the fish from sliding and falling away, once caught

IMPERIAL SHAG
Phalocrocorax atriceps
This great coastal diver has solid bones and strong swimming feet. It does not oil its plumage so that it can better submerge.

BROWN PELICAN
Pelecanus occidentalis
It stays on the shore. It uses its crop as a fishing net while it swims.

HERRING GULL
Larus argentatus
A voracious fisher and great glider. There are many species, some truly cosmopolitan.

CAPE GANNET
Morus capensis
Skillful spearfishers. They live in colonies in Africa. To help cool themselves, they have a strip of naked skin on their throats.

Salt Gland

Living in the ocean requires a few adaptations. The most notable one is the salt gland, which eliminates excess salt from the bloodstream. This way, marine birds can even drink salt water without suffering dehydration, as would be the case with humans. This gland is very efficient: it has been observed that 20 to 30 minutes after drinking a saline solution with concentrations similar to that of the ocean (4 percent), birds eliminate another solution (through the nostrils) with 5 percent salt, in the shape of water drops.

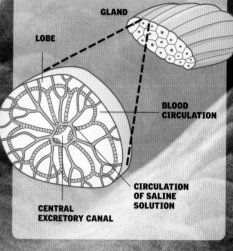

GLAND

LOBE

BLOOD CIRCULATION

CIRCULATION OF SALINE SOLUTION

CENTRAL EXCRETORY CANAL

TUBULAR NOSTRILS
Albatrosses have one on each side of their bills. On petrels and shearwaters, the tubular nostrils have merged on top of the bill, forming a single nasal tube.

BILL
It is composed of several hard plates.

DIVE FISHING

1
The bird dives to gain enough speed.

2
It folds its wings and stretches out its neck to immerse itself in the water and reach the school of fish.

3
The bird immerses its body as much as possible to catch the fish; its feathers cause it to float back up.

Fishing Methods

Many marine birds fish by diving into the sea. This way, they can access fish that swim below the surface. In order to reach deeper levels in the water, they fly up several feet, spy a school, fold their wings, and plunge with their necks stretched forward. Thanks to the buoyancy of their feathers, they are back on the surface moments later.

Freshwater Birds

This group includes birds that vary greatly—from common kingfishers to ducks to storks—and covers a wide spectrum. Freshwater birds live in rivers, lakes, and ponds for at least part of the year and are perfectly adapted to aquatic life. Some are excellent swimmers, whereas others are great divers. An important group wades in watercourses with long legs as they fish. Freshwater birds have a varied diet and are mostly omnivorous.

Ducks and Distant Cousins

The order Anseriformes includes birds that are very familiar to humans: ducks, geese, and swans, for example. They have short, webbed feet and wide, flat bills lined with lamellae (false teeth) that enable them to filter their food, catch fish, and scrape the beds of rivers and ponds. Most are omnivorous and aquatic (either staying on the surface or diving), although some species spend more time on land. They are widely distributed, and the plumage of males becomes very colorful during the courtship season.

10-13 inches
(26-33 cm)

MUSCOVY DUCK
Cairina moschata

28-34 inches
(70-85 cm)

BLACK-NECKED SWAN
Cygnus melancoryphus

26-34 inches
(66-86 cm)

WHITE-FRONTED GOOSE
Anser albifrons

FOLDED WEB

UNFOLDED WEB

HOW THEY USE THEIR FEET TO SWIM

A duck moves its feet in two ways. To advance, it spreads out its toes and uses its webbed feet to row. It closes the toes to bring the foot forward again. If the bird wants to turn, only one foot pushes to the side.

A DUCK'S DIET

1 It swims on the surface, looking for food underneath the water.

2 It sticks its head into the water, abruptly pushes back its feet, and turns its neck downward.

3 It floats facedown and pokes the bottom with its bill.

ORIFICES
Open and oval

LAMELLAE
Around the inside
edges of the bill

DUCK BILLS

are flat, wide, and slightly depressed toward the middle.
In general, their shape does not vary, but there are
species with tiny bills (the mandarin duck, for example).

2–4 inches
(5–10 cm)

1 inch
(2.7 cm)

SHOVEL-SHAPED BILL:
Typical of many ducks. The
size varies.

MANDARIN DUCK BILL:
One of the smallest-billed
species.

**FULVOUS
WHISTLING
DUCK**
*Dendrocygna
bicolor*

Wading Birds

These birds belong to an artificial order since,
from a genetic perspective, the species are
not related. They are grouped together because
adaptation to the same habitat has caused them
to develop similar shapes: long bills and necks to
perform skillful movements and thin legs designed
to wade across the water as they fish. Herons form
a special group because they are cosmopolitan and
because they have powder down feathers. Ibis and storks
also have a wide distribution (area in which they occur).
Birds that have spoon- and hammer-shaped bills are found
primarily in Africa.

IBIS (*Ibis* sp.):
Some filtrate, and others fish.

STORK (*Ciconia* sp.):
It fishes with its long bill.

SHOEBILL (*Balaeniceps rex*):
It eats among floating sedges.

HERON (*Egretta* sp.):
It fishes with its sharp bill.

COMMON SPOONBILL
(*Platalea leucorodia*): It eats
several types of aquatic animals.

HAMMERKOP (*Scopus
umbretta*): It fishes and hunts
small animals.

Divers and Other Fishers

Diving birds belong to the family Podicipediformes. They feed on small
fish and aquatic insects. They are very clumsy on the ground. In the
Coraciiformes order, common kingfishers and other similar birds find their
prey by closely watching the water. When one of these birds notices a small
fish, it spears it, catching it with its bill. In the order Charadriiformes, curlews
wander around the edges of ponds in search of food. Their long legs keep their
bodies out of the water. They are not swimmers.

12–16 inches
(30–40 cm)

16 inches
(40 cm)

7 inches
(18 cm)

GREBE
Podiceps sp.

STONE CURLEW
Burhinus oedicnemus

**COMMON
KINGFISHER**
Alcedo atthis
also known as
European Kingfisher)

THE BILL OF AN IBIS
is long and thin, ideal to
stick in the mud to look
for food.

WHITE IBIS
Edocimus albus

THE LEGS OF AN IBIS
keep the bird above the water
but close enough to fish. Ibis
also stir up the beds of lakes
and ponds.

Armed to Hunt

Birds of prey are hunters and are carnivorous by nature. They are perfectly equipped to eat living animals. Their eyesight is three times sharper than that of human beings; their ears are designed to determine the precise status of their prey; they have strong, sharp talons; and they can kill a small mammal with the pressure of their talons alone. Their hook-shaped bills can kill prey by tearing its neck with a single peck. Eagles, falcons, vultures, and owls are examples of birds of prey. Birds of prey can be diurnal or nocturnal, and they are always on the lookout.

Diurnal and Nocturnal

Eagles, falcons, and vultures are diurnal birds of prey, whereas owls are nocturnal—that is, they are active during the night. These two groups are not closely related. These birds' main prey includes small mammals, reptiles, and insects. Once they locate the victim, they glide toward it. Nocturnal birds of prey are specially adapted: their eyesight is highly developed, their eyes are oriented forward, and their hearing is sharp. The feathers on their wings are arranged in such a way that they make no noise when the bird is flying. In order to protect themselves while sleeping during the day, they have dull plumage, which helps them blend in with their surroundings.

EURASIAN EAGLE OWL
Bubo bubo
Its ears are asymmetrical and can determine the location of prey with great precision.

BALD EAGLE
Haliaeetus leucocephalus
It has a visual field of 220 degrees and a bifocal vision of 50 degrees.

CERE
Fleshy formation, somewhat thick and soft

Bills

The bills of birds of prey are hook-shaped. Some birds of prey have a tooth that works like a knife, allowing them to kill their prey, tear its skin and muscle tissues, and get to the food easily. The structure and shape of the bills of birds of prey changes, depending on the species. Scavengers (for example, vultures and condors) have weaker bills because the tissues of animals in decomposition are softer. Other species, such as falcons, catch prey with their talons and use their bills to finish it off with a violent stab to the neck, breaking its spine.

TIP
Where the tooth is located

NOSTRIL
Olfactory canals

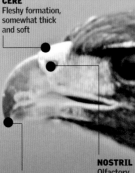

Zone-Tailed Hawk
Buteo albonotatus

BALD EAGLE
Its hook-shaped bill is common to many birds of prey.

SPARROW HAWK
Its thin bill enables it to take snails out of their shells.

FALCON
It can break the spine of its prey with its upper bill.

GOSHAWK
Its strong bill can catch prey as large as hares.

OWL PELLETS
Owls produce pellets. They swallow their prey whole and regurgitate the indigestible substances. The study of pellets makes it possible to determine the fauna of small areas with great precision.

HOW THE VULTURE HUNTS

1
Vultures feed mainly on carrion, although they are able to attack a living animal if it is vulnerable and the situation presents itself.

2
Thanks to their ability to glide on thermals, vultures can find carcasses on which to feed without wasting energy.

3
Once they find food, they must analyze the territory to know if they will be able to take flight again soon.

DIMENSIONS

The wings of birds of prey are adapted to suit their flying requirements. They can measure up to 10 feet (3 m).

CONDORS
3 to 9½ ft (0.95–2.9 m)

EAGLES
4½ to 8 ft (1.35–2.45 m)

BUZZARDS
4 to 5 ft (1.2–1.5 m)

KITES
2½ to 6½ ft (0.8–1.95 m)

RED-BACKED HAWK
3½ to 4½ ft (1.05–1.35 m)

FALCONS
2¼ to 4 ft (0.67–1.25 m)

Feet

Most birds of prey catch and kill their prey with their talons and tear away the meat with their bills. For this reason, birds' feet constitute one of the morphological characteristics of a species. The toes end in strong, sharp nails that the bird uses as pincers to catch its prey in flight. The osprey also has thorns on the soles of its feet, which help it to catch fish.

GRIFFON VULTURE
Its long toes do not have a good grasp.

FISHING EAGLE
Its toes have rough scales that look like thorns, which help it to catch fish.

GOSHAWK
It has calluses at the tips of its toes.

SPARROW HAWK
Its feet have tarsi and short, strong toes.

The Perchers Club

Passerines—or Passeriformes, the scientific name—form the widest and most diverse order of birds. What distinguishes them? Their feet are suited for perching and, therefore, for living among trees, although they can also stroll on the ground and through the brush. They inhabit terrestrial environments all over the world, from deserts to groves. Their complex sounds and songs originate from a very well-developed syrinx. Their chicks are nidicolous—that is, naked and blind at birth. In their youth, they are agile and vivacious, with very attractive, abundant, and colorful plumage.

The Smallest

Passerines are small in comparison with other birds. Their size varies from 2 inches (5 cm) bee hummingbirds (*Mellisuga helenae*) to 7 inches (19 cm) Chilean swallow (*Tachycineta leucopyga*) to 26 inches (65 cm) common raven (*Corvus corax*).

HUMMINGBIRDS
2 INCHES (5 CM)
They get so much energy from nectar that they can double their body weight by eating. However, they use this energy up during their frantic flights.

SWALLOWS
7 INCHES (19 CM)
Swallows have great agility and skill. These popular migratory birds have bodies suited for long trips.

RAVENS
26 INCHES (65 CM)
They eat everything: fruits, insects, reptiles, small mammals, and birds. They are skillful robbers of all kinds of food.

PASSERIFORMES BIRDS

Passerines have been classified into 79 families, with more than 5,400 different species.

50 %

THE PERCENTAGE OF BIRDS THAT ARE INCLUDED IN THE ORDER PASSERIFORMES

Family Album

Four basic groups have been established to facilitate the study of families: passerines with wide bills; ovenbirds, whose plumage is dull and brown (ovenbirds are noted for the great care they take in building nests); lyrebirds, whose tails have two external feathers that are longer than the others; and songbirds, with their elaborate and pleasant singing. Songbirds form the most numerous and varied group; it includes swallows, goldfinches, canaries, vireos, and ravens.

LYREBIRDS
There are only two species of these Passeriformes, and they are found only in Australia. They are very melodic and are excellent imitators of other birds. They can even imitate the sound of inanimate objects, such as horses' hooves.

SINGER

This blue-and-white swallow (*Notiochelidon cyanoleuca*) intones its pleasant and trilling chant while it flies or when it alights. Larks, goldfinches, canaries, and other passerines delight us with their trills and sounds.

HARD, SHORT BILL

The bill of a swallow is very short and tough. The swallow can use it to catch insects in flight.

SYRINX

This sound-producing organ is located at the end of the trachea. The muscles in the syrinx move the bronchial walls, which, as the air passes through, produce the melodic sounds that characterize songbirds.

SYRINGEAL CARTILAGE

TRACHEAL RING

BRONCHIAL MUSCLES

BRONCHIAL RING

LIVING AT THE EXTREMES

They range from one hemisphere to the other. They raise their chicks in the north and fly to the south to spend winter there. They fly all the way to Tierra del Fuego. Their sense of direction is remarkable. They can find and reuse their nests after returning from a migration.

A In the summer, during the reproductive season, they live in the Northern Hemisphere on the North American continent. In general, neotropical migratory birds are those that reproduce above the Tropic of Cancer.

B When winter arrives in the Northern Hemisphere, they perform a mass migration to the south, occupying the Caribbean and South America. The barn swallow travels 14,000 miles (22,000 km) during its migratory trip from the United States to southern Argentina.

PERCHING FOOT

Three toes project forward, and the well-developed hallux projects backward. This type of foot allows the bird to hold on tightly to branches.

BARN SWALLOW

(*Hirundo rustica*)
Barn swallows spend most of their time traveling to temperate zones.

WIDE BILLS

They are native to Africa and Asia and inhabit tropical zones with dense vegetation. They eat insects and fruits. They produce nonvocal sounds with the flapping of their wings. They do this during courtship, and the sound can be heard 200 feet (60 m) away.

OVENBIRDS AND THEIR RELATIVES

Their nests are completely covered structures, similar to ovens. Other members of this family build nests with leaves and straw, weaving interesting baskets. Still others dig tunnels in the ground.

3 REPTILES

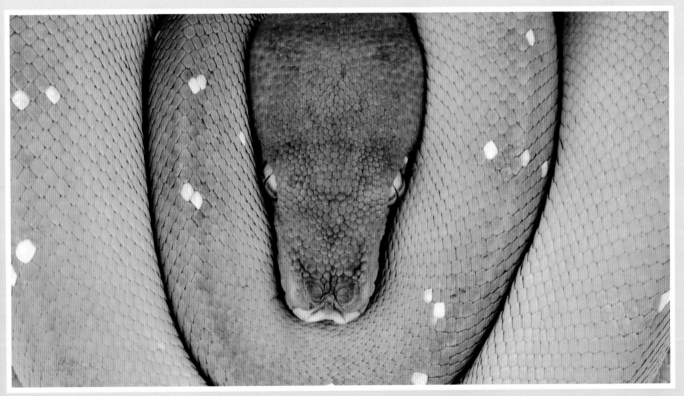

Reptiles: Background

Color plays a very important role in the life of iguanas and lizards. It helps to differentiate males and females, and when it is time to attract a mate, the members of the iguana family communicate by showing bright colors and folds of skin. Another particularity that distinguishes iguanas is their covering of epidermal scales. In addition, like all reptiles, they

GOOD VISION
Iguanas have very good vision. They see colors, and they have transparent eyelids that close easily.

are not capable of generating internal heat, so they depend on external factors to maintain their body temperature. For this reason, you will frequently see iguanas lying stretched out in the sun. When it comes to their diet, most reptiles are carnivores, with the exception of some turtles, which are herbivores. Reptiles are also characterized by their total independence from aquatic environments.

A Skin with Scales

Reptiles are vertebrates, meaning that they are animals with a spinal column. Their skin is hard, dry, and flaky. Like birds, most reptiles are born from eggs deposited on land. The offspring hatch fully formed without passing through a larval stage. The first reptiles appeared during the height of the Carboniferous period in the Paleozoic era. During the Mesozoic era, they evolved and flourished, which is why this period is also known as the age of reptiles. Only 5 of the 23 orders that existed then have living representatives today.

SOLOMON ISLAND SKINK
Corucia zebrata

EMBRIONARY MEMBRANES
They develop two: a protective amnion and a respiratory allantoid (or fetal vascular) membrane.

EYES
are almost always small. In diurnal animals, the pupil is rounded.

NICTITATING MEMBRANE
extends forward from the internal angle of the eye and covers it.

4,765 SPECIES OF LIZARDS EXIST.

Habitat

Reptiles have a great capacity to adapt, because they can occupy an incredible variety of environments. They live on every continent except Antarctica, and most countries have at least one species of terrestrial reptile. They can be found in the driest and hottest deserts, as well as the steamiest, most humid rain forests. They are especially common in the tropical and subtropical regions of Africa, Asia, Australia, and the Americas, where high temperatures and a great diversity of prey allow them to thrive.

BLACK CAIMAN
Melanosuchus niger

Crocodiles

are distinguished by their usually large size. From neck to tail, their backs are covered in rows of bony plates, which can give the impression of thorns or teeth. Crocodiles appeared toward the end of the Triassic period, and they are the closest living relatives to both dinosaurs and birds. Their hearts are divided into four chambers, their brains show a high degree of development, and the musculature of their abdomens is so developed that it resembles the gizzards of birds. The larger species are very dangerous.

OVIPAROUS
Most reptiles are oviparous (they lay eggs); however, many species of snakes and lizards are ovoviviparous (they give birth to live offspring).

THORAX AND ABDOMEN
are not separated by a diaphragm. Alligators breathe with the help of muscles on the walls of their body.

AMERICAN ALLIGATOR
Alligator mississippiensis

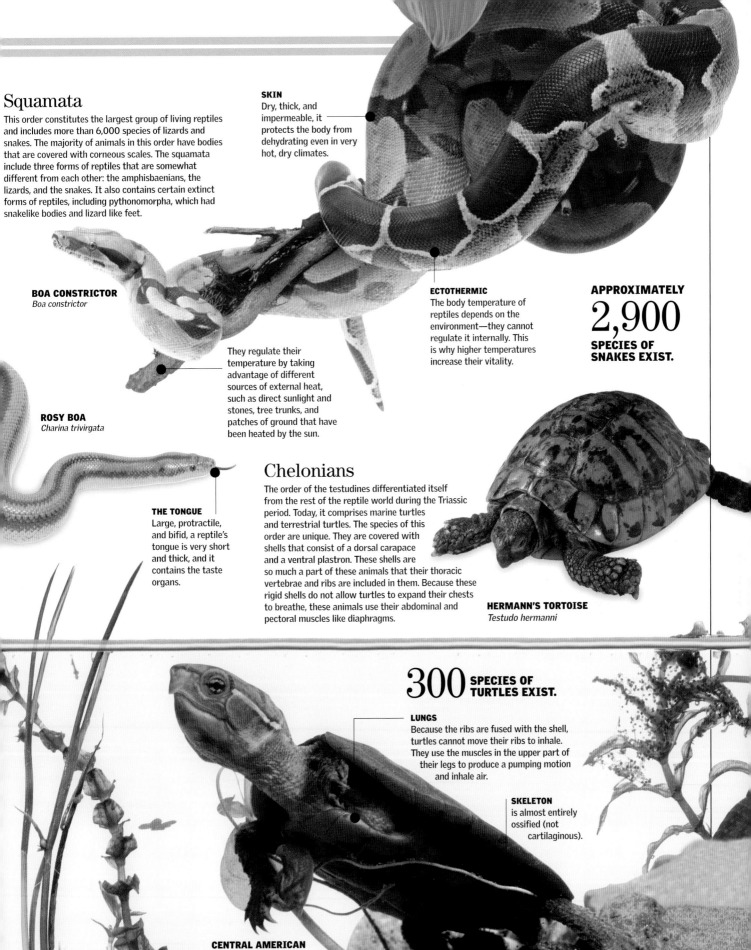

Squamata

This order constitutes the largest group of living reptiles and includes more than 6,000 species of lizards and snakes. The majority of animals in this order have bodies that are covered with corneous scales. The squamata include three forms of reptiles that are somewhat different from each other: the amphisbaenians, the lizards, and the snakes. It also contains certain extinct forms of reptiles, including pythonomorpha, which had snakelike bodies and lizard like feet.

SKIN
Dry, thick, and impermeable, it protects the body from dehydrating even in very hot, dry climates.

BOA CONSTRICTOR
Boa constrictor

ECTOTHERMIC
The body temperature of reptiles depends on the environment—they cannot regulate it internally. This is why higher temperatures increase their vitality.

APPROXIMATELY
2,900
SPECIES OF SNAKES EXIST.

They regulate their temperature by taking advantage of different sources of external heat, such as direct sunlight and stones, tree trunks, and patches of ground that have been heated by the sun.

ROSY BOA
Charina trivirgata

THE TONGUE
Large, protractile, and bifid, a reptile's tongue is very short and thick, and it contains the taste organs.

Chelonians

The order of the testudines differentiated itself from the rest of the reptile world during the Triassic period. Today, it comprises marine turtles and terrestrial turtles. The species of this order are unique. They are covered with shells that consist of a dorsal carapace and a ventral plastron. These shells are so much a part of these animals that their thoracic vertebrae and ribs are included in them. Because these rigid shells do not allow turtles to expand their chests to breathe, these animals use their abdominal and pectoral muscles like diaphragms.

HERMANN'S TORTOISE
Testudo hermanni

300 **SPECIES OF TURTLES EXIST.**

LUNGS
Because the ribs are fused with the shell, turtles cannot move their ribs to inhale. They use the muscles in the upper part of their legs to produce a pumping motion and inhale air.

SKELETON
is almost entirely ossified (not cartilaginous).

CENTRAL AMERICAN RIVER TURTLE
Dermatemys mawii

Internal Organs

The anatomy of reptiles enables them to live on land. Thanks to their dry, scaly skin and their excretion of uric acid instead of urea, they minimize water loss. The heart distributes blood in a double circuit. Crocodiles were the first vertebrates to have a four-chambered heart; the separation of the ventricles is incomplete in all other reptiles. The lungs, developed beyond those of amphibians, contribute to cardiac efficiency by allowing for greater exchange of gases.

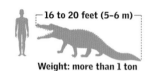

16 to 20 feet (5–6 m)

Weight: more than 1 ton

NILE CROCODILE
Crocodylus niloticus

DIET	Carnivorous
LONGEVITY	45 years in the wild and 80 years in captivity

SKIN

Reptiles have chromatophores that modify their color to a small degree. Two unique traits of crocodiles are that the skin on the head has glands that regulate the body's ionic balance, and the cloaca has glands that secrete substances crucial for mating and defense.

Jugular Region   Neck

PLACEMENT OF THE SCALES

Transverse Lines   Longitudinal Lines

Anus

The flanks and the stomach are lighter and shinier than the dorsal surface.

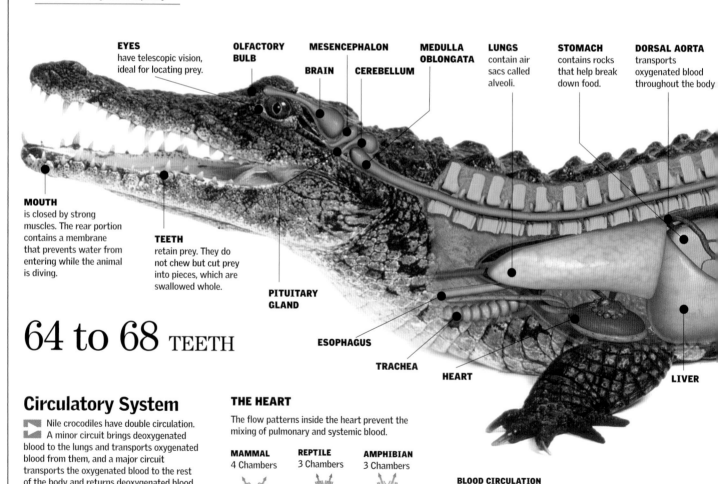

EYES have telescopic vision, ideal for locating prey.

OLFACTORY BULB

MESENCEPHALON

BRAIN   **CEREBELLUM**

MEDULLA OBLONGATA

LUNGS contain air sacs called alveoli.

STOMACH contains rocks that help break down food.

DORSAL AORTA transports oxygenated blood throughout the body

MOUTH is closed by strong muscles. The rear portion contains a membrane that prevents water from entering while the animal is diving.

TEETH retain prey. They do not chew but cut prey into pieces, which are swallowed whole.

PITUITARY GLAND

64 to 68 TEETH

ESOPHAGUS

TRACHEA   **HEART**   **LIVER**

Circulatory System

Nile crocodiles have double circulation. A minor circuit brings deoxygenated blood to the lungs and transports oxygenated blood from them, and a major circuit transports the oxygenated blood to the rest of the body and returns deoxygenated blood to the heart. Reptile hearts have two auricles (atria) and a single ventricle, which is partially divided by an incomplete ventricle partition.

THE HEART

The flow patterns inside the heart prevent the mixing of pulmonary and systemic blood.

MAMMAL 4 Chambers

REPTILE 3 Chambers

AMPHIBIAN 3 Chambers

BLOOD CIRCULATION
An ample and efficient network of blood vessels extends throughout the bodies of reptiles.

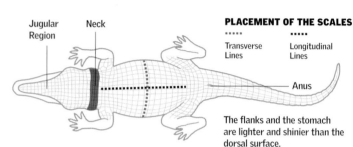

A Question of Skin

The absence of extremities and the friction produced during movement give snakes the ability to slide and to shed their skin in one motion. Other reptiles must shed their skin by tearing it off in pieces. Reptiles shed their skin regularly and continue to do so even in the last years of their lives.

NEW SKIN
is smooth and bright.

CORAL SNAKE

Micrurus altirostris
It is characterized by its smooth and brightly colored scales.

OLD SKIN
is fragile. It rips easily.

100
THE NUMBER OF TIMES A VIPER SHEDS ITS SKIN OVER ITS LIFETIME.

Melanophores

Osteoderms

Flexible Joint

GROWTH OF THE SCALES

EPIDERMIS

DERMIS

1 The dermal layer is found below the epidermis.

2 Differentiation takes place during dermal cell growth..

3 The epidermis secretes large amounts of keratin.

4 The new scales overlap each other and cover the skin.

SPLEEN

TESTICLES
Lobuled. Their ducts empty in the cloaca.

KIDNEYS
Metanephric. The ureters empty in the cloaca.

DOUBLE CAUDAL CREST

CLOACA
Shared opening of the excretory, reproductive, and digestive ducts

SIMPLE CAUDAL CREST

COLON

SMALL INTESTINE

Respiratory System

is completely pulmonary. Most reptiles possess a pair of functional lungs, with the exception of snakes, which have only one functional lung. Body-wall muscles generate the pressure differences necessary to circulate air through the airways from the nasal cavities to the pulmonary alveoli.

BREATHING

1 EXHALING
Internal organs are compressed. This, in turn, compresses the lungs and causes them to expel air.

2 INHALING
The pelvic bones rotate downward, the abdomen stretches, and the muscles cause the lungs to expand.

Abdominal Muscles

The liver compresses the lungs.

Air is expelled.

The pressure difference causes the lungs to expand and take in air.

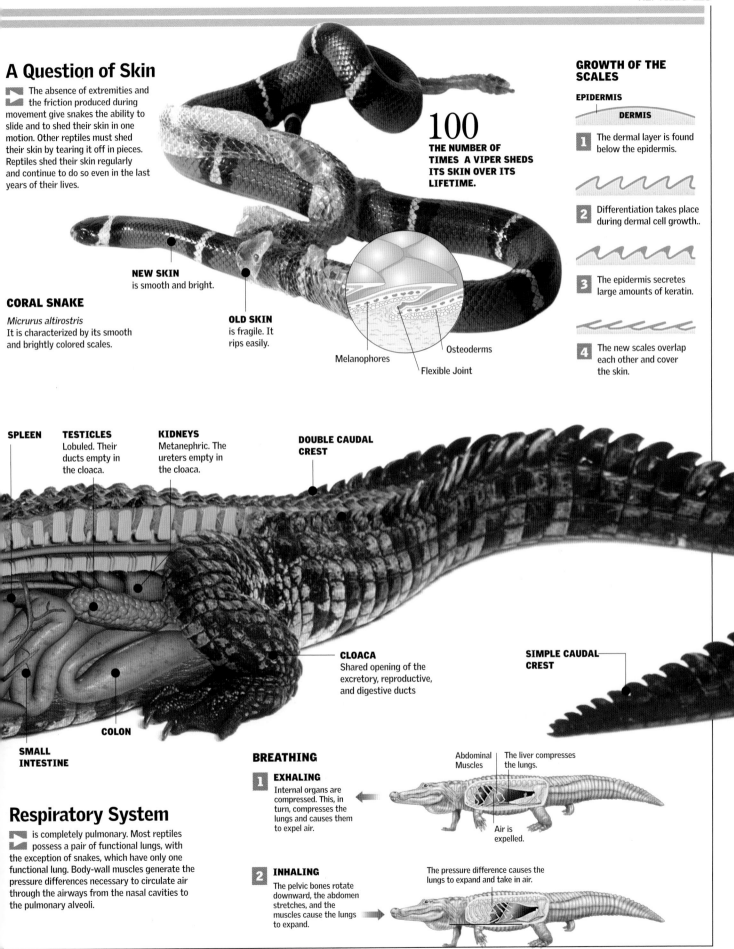

A Menu at Ground Level

Reptiles are basically carnivorous, even though some follow other food regimens. Lizards usually feed on insects. Snakes usually feed on small vertebrates such as birds, rodents, fish, amphibians, or even other reptiles. For many, the eggs of birds and other reptiles make a very succulent meal. The painted turtle is omnivorous: it eats meat and plants. Reptiles and other species are part of a larger food chain—animals eat other animals, preserving the equilibrium of the environment.

Herbivores

This diet is generally typical of other groups of animals; however, there are reptiles that feed on only green leaves and plants. The marine iguana eats only the algae that it finds under rocks on the sea floor.

GREEN IGUANA

Also called the common iguana, it is one of the few herbivorous reptiles. It feeds on green leaves as well as on some fruits.

GREEN IGUANA
Iguana iguana

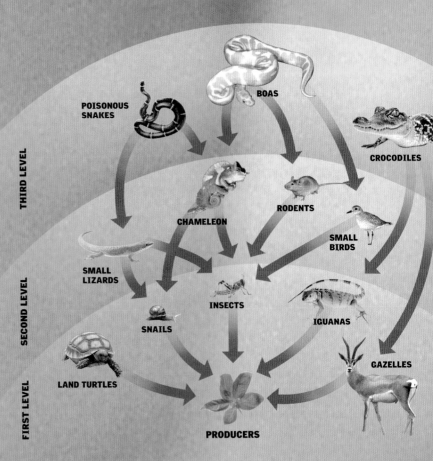

POISONOUS SNAKES

BOAS

CROCODILES

THIRD LEVEL

CHAMELEON

RODENTS

SMALL BIRDS

SECOND LEVEL

SMALL LIZARDS

INSECTS

IGUANAS

SNAILS

FIRST LEVEL

LAND TURTLES

GAZELLES

PRODUCERS

The Food Chain

Because they use photosynthesis, which permits inorganic carbon to be transformed into organic material, plants are the only true "producers" in the food chain. Herbivores feed on them and are, thus, first-level consumers. The animals that feed on the herbivores are second-level consumers, and the animals that eat other carnivores—a category that includes some reptiles—form the third level of consumers in the food chain.

Metabolism

In the case of snakes that swallow their prey whole, digestion takes weeks and sometimes even months. Their gastric juices digest even the bones of their prey.

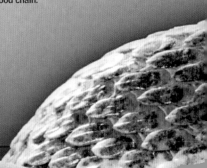

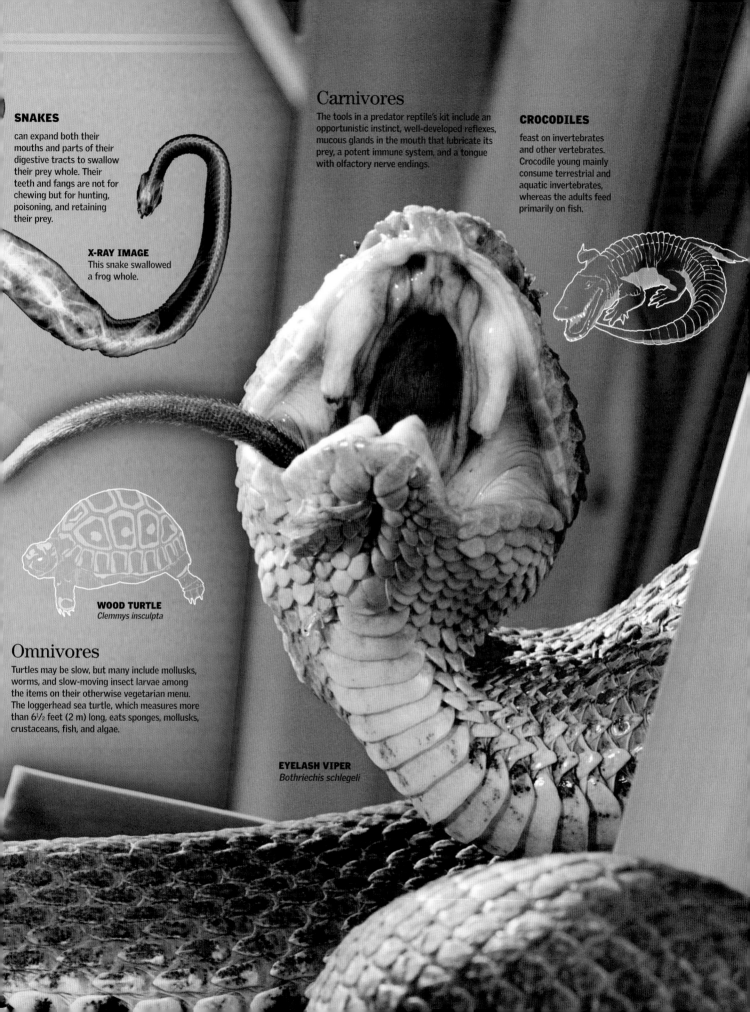

SNAKES

can expand both their mouths and parts of their digestive tracts to swallow their prey whole. Their teeth and fangs are not for chewing but for hunting, poisoning, and retaining their prey.

X-RAY IMAGE
This snake swallowed a frog whole.

Carnivores

The tools in a predator reptile's kit include an opportunistic instinct, well-developed reflexes, mucous glands in the mouth that lubricate its prey, a potent immune system, and a tongue with olfactory nerve endings.

CROCODILES

feast on invertebrates and other vertebrates. Crocodile young mainly consume terrestrial and aquatic invertebrates, whereas the adults feed primarily on fish.

WOOD TURTLE
Clemmys insculpta

Omnivores

Turtles may be slow, but many include mollusks, worms, and slow-moving insect larvae among the items on their otherwise vegetarian menu. The loggerhead sea turtle, which measures more than 6½ feet (2 m) long, eats sponges, mollusks, crustaceans, fish, and algae.

EYELASH VIPER
Bothriechis schlegeli

Reproduction

Most reptiles are oviparous. Some species lay large numbers of eggs and then allow them to develop on their own, generally in well-protected nests or hidden under dirt or sand. Marine turtles, especially green turtles, travel to the coast to lay their eggs in the sand, where they are left at the mercy of all who pass by. The females of other species, however, fiercely protect their offspring, staying near their nests for long periods of time to scare away potential predators.

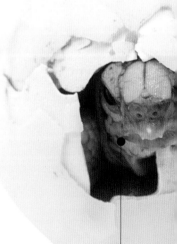

GREEN ANACONDA

Eunectes murinus
An anaconda can have more than 50 offspring, which measure nearly 3 feet (1 m) long at birth.

Eggshells

Reptile offspring develop within a liquid-filled sac called the amnion, which lies inside the egg. Most reptile eggs have soft, flexible shells, but some have much harder shells. Through the shell, the hatchling absorbs the oxygen and moisture it needs for growth, while its yolk provides it with food.

Oviparous

Reproduction that involves laying eggs in which the offspring complete their development before hatching. Some species lay large numbers of eggs and then allow them to develop on their own, generally in well-protected nests or hidden in dirt or sand. In other species, such as crocodiles, the females fiercely protect their offspring.

THE FEMALE REPRODUCTIVE SYSTEM
has two ovaries that contain the ovules, leading into two oviducts that reach the cloaca. Fertilization occurs in the forward part of the oviduct.

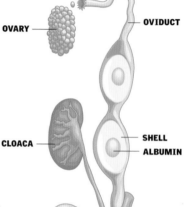

OVARY

OVIDUCT

CLOACA

SHELL

ALBUMIN

CLOACA

1

Growth

The egg is buried by the mother, and the embryo begins to develop. The egg provides the necessary oxygen and food.

2

Fracture

The pressure exerted on the shell from the movements of the animal within such a cramped space causes the shell to break from the inside.

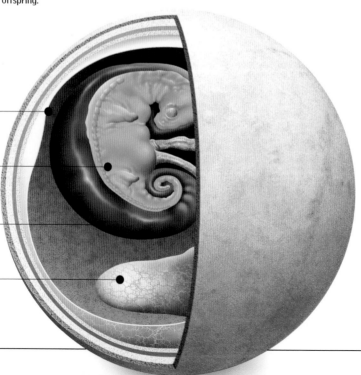

LEOPARD TORTOISE

SHELL
allows oxygen to enter, so the embryo can breathe.

EMBRYO
Protected from drying out, it can survive without water.

YOLK SAC
surrounds the embryo and stores food for its birth.

ALLANTOIS
Prolongation of the embryonic intestines

EGG TOOTH
A corneous, or horny, spine on the beak for breaking the shell during hatching

Ovoviviparous

The eggs remain in the mother's body, and hatching occurs there. The hatched young are diminutive versions of adult animals. Already independent, they do not receive any care from their parents.

145 to 160 days

IS THE INCUBATION PERIOD OF THE LEOPARD TORTOISE.

Exit

It can take the baby animal an entire day to exit, and it will have a small sac hanging from its navel. This is the sac of yolk that provided it with food while it was incubating.

MOUTH
is the first part to appear.

CARAPACE
Its growth makes the egg break.

FOOT
already has mobility, thus allowing the baby reptile to walk.

CARAPACE (SHELL)
is already completely formed at birth.

Hatching

The turtle is ready to come out and starts to break the egg with its body. Hatching occurs.

LEOPARD TORTOISE
Geochelone pardalis

HABITAT	Africa
DIET	Herbivorous
SIZE	23–26 inches (60–65 cm)
WEIGHT	77 pounds (35 kg)

FER-DE-LANCE

Bothrops atrox
In one litter, it can produce up to 80 offspring, each with a length of 13 inches (34 cm).

CONSISTENCY OF THE EGGS
The eggshell can be soft or hard. Soft eggshells are usually found in lizards and snakes, whereas hard eggshells are common in turtles and crocodiles.

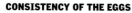

HARD   **SOFT**

Viviparous

As is the case with most mammals, the whole embryonic developmental cycle occurs inside the mother's body, and the embryo obtains food from close contact with maternal tissues.

Crocodiles, Lizards, Turtles and Snakes

Because of their long, powerful bodies and sharp teeth, crocodiles are among the most dangerous predators. When they are small, they eat small fish, frogs, and insects. When fully grown, however, they can devour large animals and even humans. Did you know that lizards are the most numerous reptiles in the world today? This group includes a wide variety of species

GREEN TREE PYTHON
This tree-dwelling green python usually coils around a branch and waits with its head hanging down, ready to attack. It eats small mammals and birds.

of all shapes and sizes. Discover the amazing world of snakes, what they are like inside, where they live, and how they hunt their prey, as well as why some eat only eggs and others, have to suffocate their prey by coiling around it. Although people may think that turtles are peaceful creatures, many are actually carnivorous hunters that eat small invertebrates, fish, and even some larger animals.

Lizards

izards are the largest group of reptiles. They live in most environments except for extremely cold regions, because they cannot regulate their own body temperatures. There are land-dwelling, underground, tree-dwelling, and even semiaquatic lizards. They can walk, climb, dig, run, and even glide. Lizards often have differentiated heads, movable eyelids, a rigid lower jaw, four five-toed feet, a long body covered with scales, and a long tail. Some can even shed their tails when threatened.

DAY GECKOS
Phelsuma sp.

STICKY TOES

Chameleons

live in Africa, especially in southeastern regions and on Madagascar. They live in forests, where they use their prehensile tails and toes to climb trees. Their well-known ability to change color is important when they face danger or when they begin to court.

Camouflage

is an adaptive advantage. By blending in with the vegetation surrounding them, lizards can escape the notice of both their predators and their prey.

LIFESAVING RECOURSE
Between each vertebra, there are rupture planes enabling the tail to separate from the body.

AUTOTOMIC TAIL
Certain lizards can shed their tails many times during their lives. In dangerous situations, they may even shed it voluntarily in order to flee their confused predators. Later the tail grows back.

TELESCOPIC EYES

Geckos and Skinks

are lizardlike animals of the family Gekkonidae that live in warm regions. Their limbs are very small. (In fact, some species have none at all!) Their bodies are covered with smooth, shiny scales.

MELLER'S CHAMELEON
Chamaeleo melleri

TAIL
curls up when necessary.

SKIN
has cells with many pigments.

PREHENSILE TOES
can surround a branch and hold on tight.

CLAW

4,765
LIZARD SPECIES EXIST IN THE WORLD.

Heloderma

comprise only two species, which live in the United States and Mexico. They feed on invertebrates and small vertebrates. Their bodies are massive, and their skin is covered with small knobs. They are the only poisonous lizards, and their bite can be dangerous to humans.

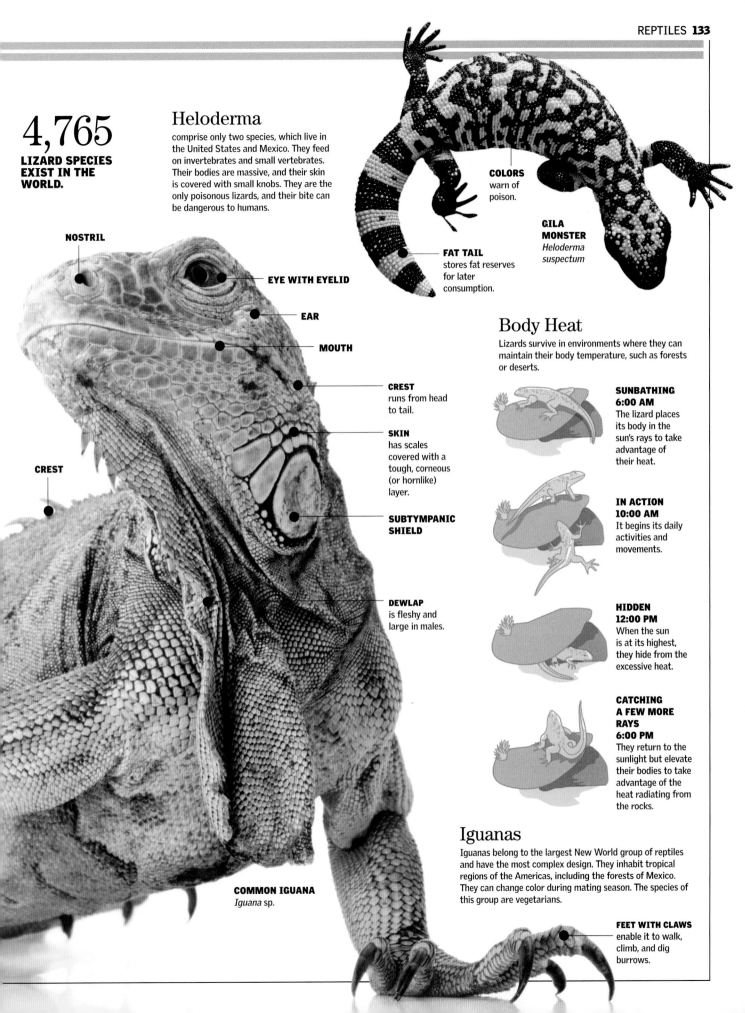

COLORS
warn of poison.

GILA MONSTER
Heloderma suspectum

FAT TAIL
stores fat reserves for later consumption.

NOSTRIL

EYE WITH EYELID

EAR

MOUTH

CREST
runs from head to tail.

SKIN
has scales covered with a tough, corneous (or hornlike) layer.

SUBTYMPANIC SHIELD

CREST

DEWLAP
is fleshy and large in males.

COMMON IGUANA
Iguana sp.

Body Heat

Lizards survive in environments where they can maintain their body temperature, such as forests or deserts.

SUNBATHING 6:00 AM
The lizard places its body in the sun's rays to take advantage of their heat.

IN ACTION 10:00 AM
It begins its daily activities and movements.

HIDDEN 12:00 PM
When the sun is at its highest, they hide from the excessive heat.

CATCHING A FEW MORE RAYS 6:00 PM
They return to the sunlight but elevate their bodies to take advantage of the heat radiating from the rocks.

Iguanas

Iguanas belong to the largest New World group of reptiles and have the most complex design. They inhabit tropical regions of the Americas, including the forests of Mexico. They can change color during mating season. The species of this group are vegetarians.

FEET WITH CLAWS
enable it to walk, climb, and dig burrows.

Komodo Dragon

This animal is the largest lizard in the world. It is related to monitor lizards and can grow up to 10 feet (3 m) long and weigh up to approximately 330 pounds (150 kg). These endangered lizards live only on a group of islands in Indonesia. They are carnivorous and are known for their ferocity in attacking their prey. Their saliva is full of bacteria that can kill their prey with only one bite. They can detect other Komodo dragons from several miles away.

INDONESIA

KOMODO NATIONAL PARK

BANTA

SUMBAWA

PADAR

FLORES

KOMODO

RINCA   KODE MONTONG

KOMODO DRAGON
Varanus komodoensis

HABITAT (APPROX.)	900 square miles (2,300 sq km)
NUMBER OF DRAGONS	Less than 5,000

TOUGH SKIN

is covered with black, brown, or dark gray scales.

CLAWS

Its five claws are very sharp. It uses them to hold onto its dying prey.

SIZE AND WEIGHT

Males can grow more than 10 feet long. Females are somewhat smaller.

Weight
330 pounds (150 kg)

10 feet (3 m)

KOMODO DRAGON

Weight
22 pounds (10 kg)

3⅓ feet (1 m)

IGUANA

6 feet
(1.8 m)

Weight
175 pounds
(80 kg)

HUMAN

STOMACH

Like most reptiles, Komodo dragons have a stomach that can expand enormously. This enables them to gulp down up to 70 percent of their own weight in a single meal.

5,000
lizards
**OF THE FAMILY VARANIDAE
LIVE IN THE WILD ON
SIX SMALL INDONESIAN
ISLANDS, INCLUDING
KOMODO ISLAND.**

A Long Hunt

Komodo dragons have an acute sense of smell that can detect the presence of other animals up to 2 miles (3 km) away. They track their prey using their forked tongues to detect scents from molecules in the air. Jacobson's organ, located inside the mouth, helps the lizard to locate its prey more rapidly and consume less energy while tracking it.

How It Attacks Its Prey

1 SEARCH
The dragon searches for food with its forked tongue. When chasing its prey, it can reach speeds of up to 11 miles (18 km) per hour.

2 BITE
Following the scent, the dragon captures its prey, which dies after being bitten. Its favorite prey is deer and wild boars.

3 FEEDING
The dragon feeds rapidly, using the flexible joints of its jaws and skull. It digests not only chunks of meat but also the skin and bones of its prey.

4 STRUGGLE
Smelling a meal, more dragons approach. The largest get the best portions. The younger ones keep their distance, since the adults may act as cannibals.

SMELL

They have an acute sense of smell that can detect the odor of decomposing flesh at a distance of up to 3 miles (5 km).

SALIVA

contains bacteria that are harmful to its prey. Antibacterial substances in the Komodo dragon's blood protect it from their harmful effects.

TONGUE

is forked and is used for tasting, smelling, and feeling. It can perceive various airborne particles, helping it detect prey.

Deadly Saliva

The saliva of Komodo dragons is full of bacteria that can quickly kill its prey by causing septicemia. To kill its prey, the Komodo dragon needs to bite it only once. An analysis of its saliva revealed 60 types of bacteria, 54 of which cause infection. These bacteria are known to cause putrefaction of dead animals, including the bacteria *Pasteurella multocida* (one of the deadliest), *Streptococcus*, *Staphylococcus*, *Pseudomonas*, and Klebsiella. When combined, they are a deadly weapon.

PASTEURELLA MULTOCIDA
Bacteria that affects the gastrointestinal and respiratory tracts of mammals and birds

Changing Colors

Chameleons are well known for their ability to change color. Another interesting fact is that their tongue can stretch great distances in seconds. They live mostly in Africa. Their prehensile tails and toes make them excellent climbers. Another helpful characteristic is that their eyes can move independently of one another, providing them a 360 degree field of vision. Their flat bodies help them to balance and to hide among the leaves.

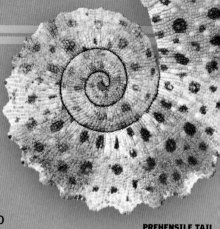

PREHENSILE TAIL
They use their long curved tails to hold on to branches without using their feet.

BONE
acts as a support for discharging the tongue.

Protractible Tongue

Long and lightweight, the chameleon's tongue is sticky and can be extended. Chameleons throw their tongues outward like projectiles to hunt their prey.

1 **CONTRACTION**

Several sheets of collagen between the tongue and the accelerating muscle are compressed in the form of a spiral, which stores the energy necessary to propel the tongue outward.

How It Changes Color

The color-changing ability of chameleons, well known in popular sayings and songs, is not an adaptation to the environment as is widely believed. Rather it is related to changes in light and temperature, courtship behaviors, or the presence of a predator. Color changes are caused by the action of hormones on pigment cells in the skin. These specialized cells, located in each layer of the dermis, react and change color, camouflaging the chameleon from its predators.

A When the upper layer (chromatophores) detects a yellow color, the blue light of the guanophores (white chromatophores) becomes greenish.

PIGMENT CELLS

Reflected light
Incident light

Chromatophores

Guanophores

Melanophores

PANTHER CHAMELEON
Furcifer pardalis

RANGE	Madagascar
HABITAT	Coastal Regions
LIFESTYLE	Diurnal

|←— 14–20 inches (35–50 cm) —→|

Feeding Habits

These diurnal hunters wait for victims to pass by. Their diet includes arthropods and small invertebrates. Among insects, they prefer crickets, grubs, cockroaches, and moths. Other species on the menu also include songbirds and mice.

Up to 600%
OF THE TONGUE'S LENGTH IS CURLED UP IN RESERVE.

TONGUE
Covered with collagen tissue.

TIP
The tip of the tongue spreads out and captures the prey with its sticky surface.

2 UNFOLDING
The accelerating muscle compresses the energy-storing collagen tissues, launching the tongue toward its target.

3 RETRACTION
When the elastic tissues contract again, they roll up the tongue and return it to its initial position with the prey sticking to it.

B The melanophores contain a dark pigment called melanin, which regulates the brightness and the amount of light reflected, varying its color.

Reflected light   Incident light

FEET
The toes are divided into two parts, with two toes on the outside and three on the inside.

2 TOES

3 TOES

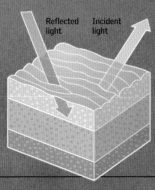

Venerated and Feared

Crocodiles—along with their relatives, the alligators, caimans, and gavials—are very ancient animals. They belong to the same group that included the dinosaurs and have changed very little in the last 65 million years. They can go for long periods without moving; during these times, they sun themselves or rest in the water. However, they can also swim, jump, and even run at high speed to attack with force and precision. In spite of their ferocity, female crocodiles provide more care for their young than any other living group of reptiles.

LOWER JAW
The lower teeth are invisible when the mouth is closed.

SCALES are flat on the tail.

GAVIAL
Gavialis gangeticus

HABITAT	Freshwater
NUMBER OF TYPES	One
DEGREE OF DANGER	Harmless

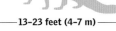

—13–23 feet (4–7 m)—

GAVIAL
has a long, narrow snout, with long front teeth.

CROCODILE
has a V-shaped snout, narrower than the alligator's.

ALLIGATOR
has a wide, short, U-shaped nose.

The Gavial

is the strangest of all crocodiles. Its long, narrow snout with small, sharp teeth sweeps through the water. Its interlocked, outward-curving teeth are perfect for catching slippery fish. Adult males drive away their rivals with loud buzzing sounds that they make by exhaling air through a bump on their noses.

SNOUT
Long, narrow nose

TEETH
are longest in front.

1 It moves forward with its four limbs.

The front legs begin the movement.

2 Its legs are suspended.

Then the hind legs come into action.

3 The cycle starts over.

The tail is raised to avoid acting as a brake.

9 miles per hour (15 km/h)
IS THE SPEED THEY CAN REACH AT A FULL RUN.

CLAWS

ALLIGATOR
Alligatoridae

HABITAT	Freshwater
NUMBER OF TYPES	Eight
DIET	Insects, Mammals, Birds

10 to 20 feet (3–6 m)

SCALES

Alligators and Caimans

Alligators and caimans are almost completely limited to freshwater. They make their nests by piling up grass, dirt, and leaves to lay their hard-shelled eggs. The female often remains near the nest to keep would-be thieves from invading. Although they look clumsy, alligators can use their jaws with precision. The female often helps her eggs to hatch by putting them in her mouth, where she rolls them against her palate with her tongue until they crack.

JOINT

NILE CROCODILE
Crocodylos niloticus

HABITAT	Freshwater and Saltwater
NUMBER OF TYPES	One
LIFE SPAN	70 years

16 to 20 feet (5–6 m)

TEETH
number from 64 to 68. The fourth tooth on the lower jaw is visible when the mouth is closed.

How They Move

Although their preferred form of movement is swimming or crawling, crocodiles can run for short distances if they feel threatened. They can reach speeds of up to 9 miles per hour (15 km/h) when running, with their abdomens supported above their knees and their elbows slightly bent. They can go even faster when sliding on mud.

30 DEGREES

POSTURE

Semicrouched. The knees and elbows are slightly bent.

SWIMMING

Using its tail for locomotion, it moves with agility through the water.

Crocodiles

have four feet. In this way, they are very similar to lizards. They are distinguished by their great size and ferocity. Several rows of bony plates that look like spines or teeth run down the length of their back. They can stay in the water for long periods of time, and they are able to swallow underwater without drowning. They make their nests in holes on the beach. The Johnston's, or freshwater, crocodile, of tropical northern Australia, can gallop to the water by raising all four feet off the ground.

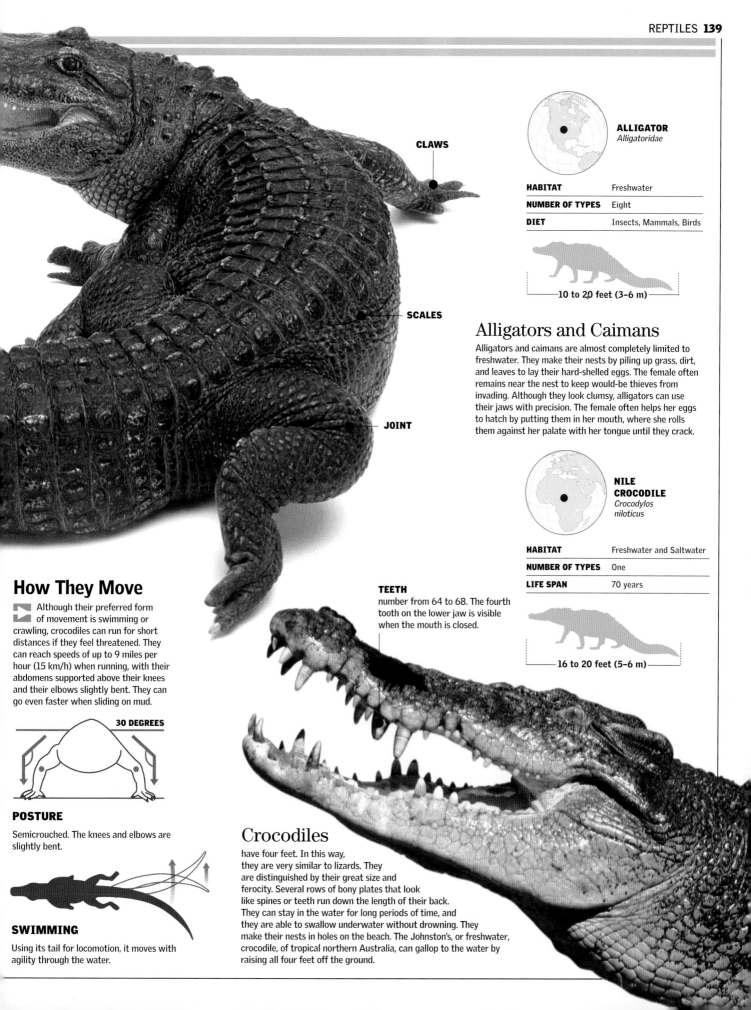

Slow but Steady

Since their appearance on the earth about 230 million years ago, turtles have changed very little. Turtles can live on land, in fresh water, or in salt water. However, they all need light and heat to survive, and they all lay their eggs on land. Although aquatic turtles are nearly all carnivorous, some land-dwelling species are herbivorous. Turtles' most noticeable trait is their hard shell, which encloses and protects the soft part of their bodies. It also camouflages them for protection from predator species.

Freshwater Turtles

Most turtle species live in fresh water. They are distinguished by their feet, which are partially or totally webbed and are used for swimming, an activity at which they are highly skilled. They can also be identified by their shells, which are flatter than those of land-dwelling turtles. Some freshwater turtles are quite well adapted to living on land. In general, they prefer warm climates with abundant vegetation, so they commonly live near swamps and rivers located in subtropical areas around the world. Their shells may have particular characteristics, depending on the species. The American box turtle, for example, can completely close its shell.

APPROXIMATELY

300–350

TURTLE SPECIES EXIST.

HEAD
has a pointed nose.

NECK
Much longer than that of other species

TYPES OF SHELLS
Turtle shells differ according to each species' habitat.

STREAMLINED
Leatherback Turtle

FLAT
Red-Eared Slider

CRESTED
Alligator Snapping Turtle

SHELL
This species has a very soft, thin shell.

CHINESE SOFT-SHELLED TURTLE

Pelodiscus sinensis
They live in swamps and streams. Their diet consists of fish and mollusks.

Ocean Species

are the rarest. They live in warm waters and are excellent swimmers. They have flippers instead of feet. The front flippers move them forward, and the back flippers act as a rudder for steering. Their shells of oceanic turtles are flattened into a streamlined shape. These turtles have developed a dual respiratory system that allows them to remain submerged for up to two hours.

SHELL
Small, flat, and joined to the skeleton

HAWKSBILL TURTLE

Eretmochelys imbricata
Sea turtles are usually heavy and large. This Caribbean species can weigh up to 141 pounds (64 kg).

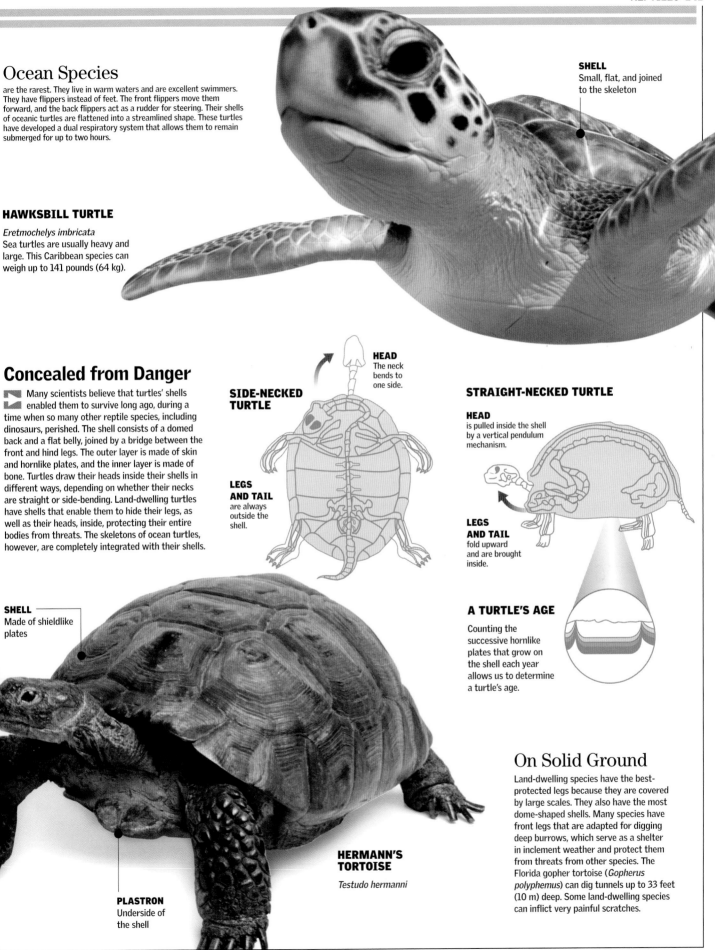

Concealed from Danger

Many scientists believe that turtles' shells enabled them to survive long ago, during a time when so many other reptile species, including dinosaurs, perished. The shell consists of a domed back and a flat belly, joined by a bridge between the front and hind legs. The outer layer is made of skin and hornlike plates, and the inner layer is made of bone. Turtles draw their heads inside their shells in different ways, depending on whether their necks are straight or side-bending. Land-dwelling turtles have shells that enable them to hide their legs, as well as their heads, inside, protecting their entire bodies from threats. The skeletons of ocean turtles, however, are completely integrated with their shells.

SIDE-NECKED TURTLE

HEAD
The neck bends to one side.

LEGS AND TAIL
are always outside the shell.

STRAIGHT-NECKED TURTLE

HEAD
is pulled inside the shell by a vertical pendulum mechanism.

LEGS AND TAIL
fold upward and are brought inside.

A TURTLE'S AGE

Counting the successive hornlike plates that grow on the shell each year allows us to determine a turtle's age.

SHELL
Made of shieldlike plates

PLASTRON
Underside of the shell

HERMANN'S TORTOISE

Testudo hermanni

On Solid Ground

Land-dwelling species have the best-protected legs because they are covered by large scales. They also have the most dome-shaped shells. Many species have front legs that are adapted for digging deep burrows, which serve as a shelter in inclement weather and protect them from threats from other species. The Florida gopher tortoise (*Gopherus polyphemus*) can dig tunnels up to 33 feet (10 m) deep. Some land-dwelling species can inflict very painful scratches.

Internal Structure

Snakes are scaly reptiles with long bodies and no legs. Some are poisonous, but others are not. Like all reptiles, they have a spinal column and a skeletal structure composed of a system of vertebrae. The anatomical differences between species reveal information about their habitats and diets—climbing snakes are long and thin, burrowing snakes are shorter and thicker, and sea snakes have flat tails that they use as fins.

COLD-BLOODED
Their temperature varies according to the environment. They do not generate their own body heat.

HEART
The ventricle has an incomplete partition.

ESOPHAGUS

LUNG

EMERALD TREE BOA
Corallus caninus

LARGE INTESTINE

TREE BRANCH
Boas can change color to imitate the branch they are curled around.

Primitive Snakes

Boas and pythons were the first snake species to appear on the earth. Many have claws or spurs as vestiges of ancient limbs of their ancestors. They are not poisonous, but they are the largest and strongest snakes. They live in trees, and some, such as the anaconda—a South American boa—live in rivers.

33 feet
(10 m)
LENGTH OF A PYTHON

THE SPINAL COLUMN
is composed of an assembly of jointed vertebrae with prolongations that protect the nerves and arteries. The system makes them enormously flexible.

VERTEBRAE

NEURAL ARCH

BODY OF THE VERTEBRA

HEMAL KEEL

SPOTTED PYTHON

Antaresia maculosa inhabits the forests of Australia.

FLOATING RIBS
allow the body to increase in size.

VERTEBRA

FLOATING RIB

RANGE OF MOTION OF THE RIBS

400 vertebrae THE NUMBER A SNAKE CAN HAVE

LIVER
is long and located along the esophagus.

BLADDER

STOMACH

SPLEEN

SCALES
are generally found in the dorsal region.

IDENTIFICATION OF SOME POISONOUS AND NONPOISONOUS VIPERS

POISONOUS

HEAD
Typically wide and triangular

BODY
Relatively long and rough

TAIL
Suddenly narrows (like a rattle) but does not end in a point

NONPOISONOUS

HEAD
Typically narrow; hard to distinguish from the neck

BODY
Narrow, with smooth scales

TAIL
Narrows gradually and ends in a point

TYPES OF MOVEMENT DEPENDING ON HABITAT

RECTILINEAR
Rainbow Boa

SERPENTINE
King Cobra

SIDEWINDING
Desert Snakes

CONCERTINA
Rattlesnake

SMALL INTESTINE
is divided into a small tract and a large tract, which ends well before the tip of the tail.

SKIN
Many species of snake have no scales on the underside.

OVARIES
The female reproductive organs

INFRARED PITS

Snakes of the family Viperidae are distinguished by two thermoreceptive pits on each side of their heads, which enable them to sense differences in temperature. Some pits are extremely sensitive, helping the snake to gauge the size of its prey when it hunts at night.

"Blind" Snakes

Some subtropical and tropical snake species live underground and only come out in droughts or floods. These are the smallest snakes; some are no longer than 4 inches (10 cm). They have large heads, few teeth, and bodies covered in very soft, slippery scales, which enable them to slide into anthills and termite hills, their only sources of food. Their eyes, which are covered with scales, barely work.

Sophisticated Snakes

Snakes of the family Viperidae, as well as other poisonous snakes that appeared later, have highly acute senses and a mouth apparatus with a system of retractable fangs for injecting venom.

REPRODUCTION
is sexual, and most species lay eggs. Some species give birth to live young.

2,978
SNAKE SPECIES EXIST.

GABOON VIPER

Bitis gabonica

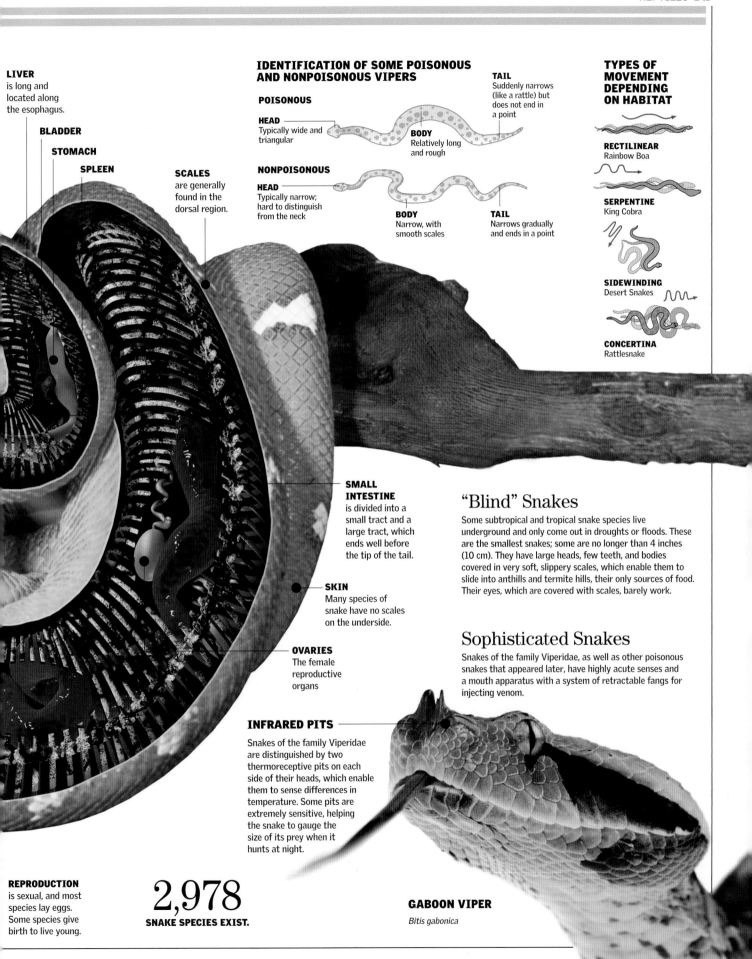

Deadly Embrace

Snakes have developed a wide range of techniques to kill their prey. For example, both boas and pythons are powerful constrictors, meaning that they kill by asphyxiating their prey rather than poisoning them with venom. Although boas and pythons belong to the same category of snakes (a category that includes the largest species in the world, the famous anaconda and reticulated python of Africa and Asia), their reproductive systems differ from one another. Their large size makes them heavy and slow moving, so they are easy prey for hunters, who kill them for their hides and meat.

JAWS
Snakes of the boa family (Boidae) have supraorbital and premaxillary bones.

TEETH
interlock.

SCALES
Heat-sensitive

Curved teeth

Smaller to larger

Flexible ligament

1 Teeth

The snake seeks out the head of its victim so that its prey will not be able to fight back. It takes the prey with its curving front teeth, keeping its victim from escaping. This enables it to suffocate its prey by coiling itself around the prey's body and squeezing (constricting).

AMAZON TREE BOA
Corallus hortulanus

RANGE South America

HABITAT Trees

LENGTH 7 feet (2 m)

7 feet (2 m)

Tree Boa

can measure up to 7 feet (2 m) long and lives in trees. Its color blends in with the surrounding foliage, concealing it from predatory birds. Its prehensile tail holds firmly onto branches, while its head hangs down so that it can pounce on passing birds or mammals.

2 Constriction

The snake seeks the head of the prey to keep it from attacking. Its entire body surrounds its prey in a deadly embrace. Every time the victim breathes, the squeeze tightens. The prey dies of suffocation.

EPAXIAL MUSCLES RELAXED

Relaxed Epaxial muscles

EPAXIAL MUSCLES CONTRACTED

Contracted epaxial muscles

Constricting ring formation

Spinal column

3 Widest Opening

When the prey is dead, the snake releases its grip and begins to feed. It swallows the head first and gradually releases the rest of the body from its coils. The time it takes to eat the prey depends on the animal's size and ranges from a few minutes to one or two hours.

The skin stretches, and the scales separate.

33 feet
(10 m)

THE LENGTH TO WHICH A GREEN ANACONDA (EUNECTES MURINUS) CAN GROW

The dead prey is ingested using the trunk muscles, which move the prey through the snake's body.

Ovoviviparous
THIS IS HOW CONSTRICTORS REPRODUCE.

A Specialized Mouth

The most primitive snakes have heavy skulls and few teeth. Most snakes, however, have lighter skulls and jointed jawbones. These joints are loose and can easily become dislocated so that the snake can swallow prey larger than the natural shape of its own head. The teeth are fixed in the upper jaw or the palate, and the fangs for injecting venom may be located either at the front or at the back of the mouth. Some species, as well as being large and powerful, have retractable fangs, allowing them to close their mouths when their fangs are not in use.

Cranial Anatomy

is directly related to each species' diet and—in the case of venomous snakes—to its system for injecting poison. Most snakes have small skulls with jawbones that can be separated voluntarily by sliding them along a kind of perpendicular rail, which consists of a bone called the quadrate. This greatly increases the size of the snake's mouth.

① Viperid

This type of skull has small teeth and large, retractable fangs that are thick or hooked.

JACOBSON'S ORGAN

gives the snake an excellent sense of smell. It consists of two cavities in the palate, into which the snake brings its tongue after "tasting" the outside air. That is why snakes continually stick out their tongues.

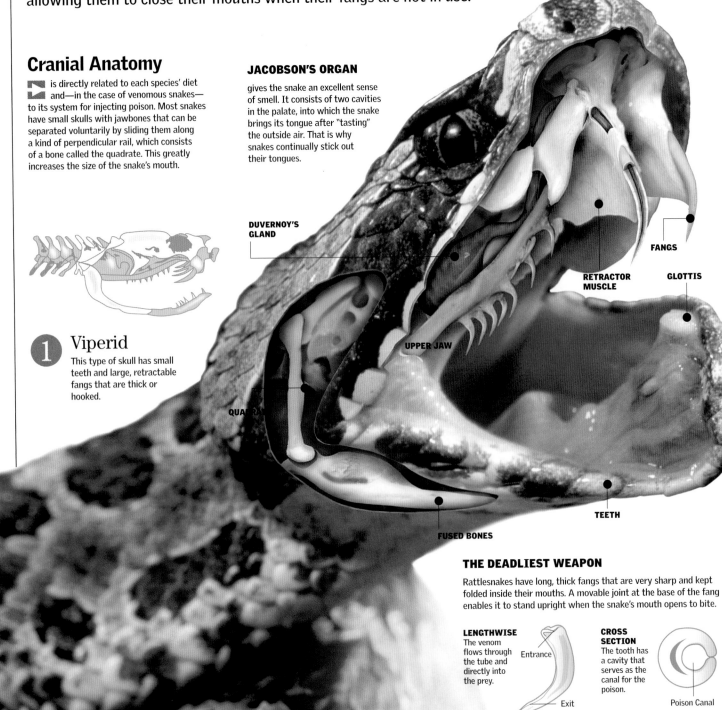

DUVERNOY'S GLAND

FANGS

RETRACTOR MUSCLE

GLOTTIS

UPPER JAW

QUADRATE

TEETH

FUSED BONES

THE DEADLIEST WEAPON

Rattlesnakes have long, thick fangs that are very sharp and kept folded inside their mouths. A movable joint at the base of the fang enables it to stand upright when the snake's mouth opens to bite.

LENGTHWISE
The venom flows through the tube and directly into the prey.

Entrance

Exit

CROSS SECTION
The tooth has a cavity that serves as the canal for the poison.

Poison Canal

Primitive Snakes

Boas and pythons are called primitive, because they have neither fangs nor venom. These snakes have several rows of small, inward-curving teeth used for holding prey and swallowing quickly without letting the prey worm out. This feature is necessary for the snakes to be able to hold onto their prey, because they lack venom. Venomous snakes, on the other hand, have little need to worry about their prey escaping, because they know that, after it is injected with the toxic substance, it will not be able to get very far.

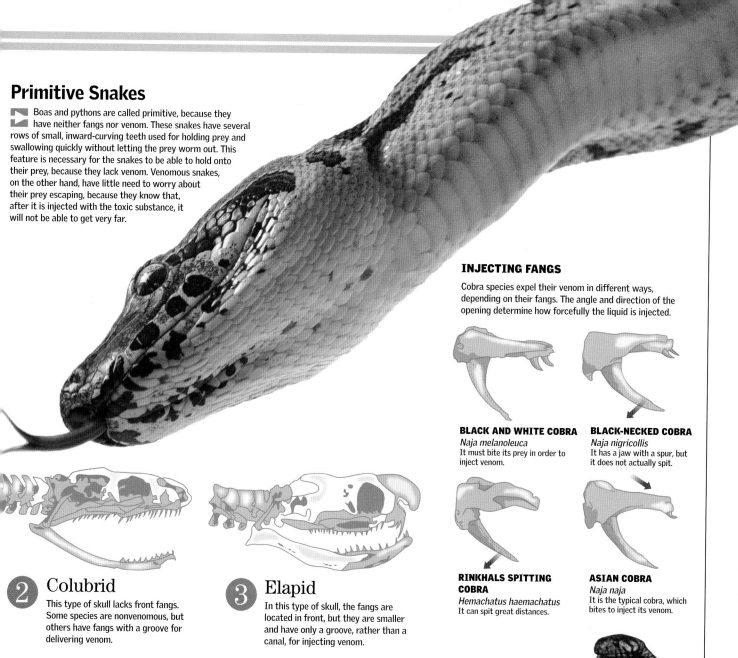

INJECTING FANGS

Cobra species expel their venom in different ways, depending on their fangs. The angle and direction of the opening determine how forcefully the liquid is injected.

BLACK AND WHITE COBRA
Naja melanoleuca
It must bite its prey in order to inject venom.

BLACK-NECKED COBRA
Naja nigricollis
It has a jaw with a spur, but it does not actually spit.

RINKHALS SPITTING COBRA
Hemachatus haemachatus
It can spit great distances.

ASIAN COBRA
Naja naja
It is the typical cobra, which bites to inject its venom.

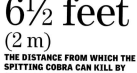

② Colubrid

This type of skull lacks front fangs. Some species are nonvenomous, but others have fangs with a groove for delivering venom.

③ Elapid

In this type of skull, the fangs are located in front, but they are smaller and have only a groove, rather than a canal, for injecting venom.

Venom System

consists of the two Duvernoy's glands, one on each side of the skull, which produce venom and are connected to the fangs. When biting, muscle contractions exert pressure on the gland and activate the injection mechanism.

6½ feet
(2 m)
THE DISTANCE FROM WHICH THE SPITTING COBRA CAN KILL BY SPRAYING ITS VENOM

Spitting Venom

Forty cobra species can spray their venom from a distance. They spray in self-defense when they feel threatened. They can direct the stream into an enemy's eyes, causing grave damage or even death. The shape of their fangs is essential to this defense.

NONSPITTING
The long canal points downward and has a beveled edge at the end. The stream loses momentum.

SPITTING
The canal's opening points forward and is narrow so that it can eject the venom more forcefully.

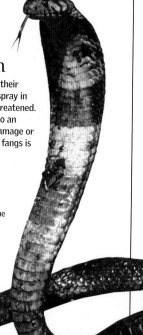

A SOLENOGLYPHS
The hollow fangs are the only teeth in the jaw. They are long and retractable and inject venom into the prey's tissues.

B PROTEROGLYPHS
Small fangs in the front of the jaw, fixed in position and with a rear groove for conducting venom

C OPISTHOGLYPHS
Fangs in the back, with no canal or groove. The prey must be held in place.

Cobras

Are an important group of snakes in the family Elapidae. Easily recognized by their outspread hoods, they are well known worldwide, mostly because of their use by snake charmers. Many cobra species carry deadly venom. Some can even spit from several yards away. Cobras of the *Naja* genus are the most widely recognized. They are widespread in Asia and were only recently recognized as ten separate species. All are predatory; many eat only snakes.

SMOOTH
Cobra species have smooth scales.

RED SPITTING COBRA
Naja pallida
One of 40 species of spitting cobras, it inhabits the Horn of Africa, where it is widely feared. It is distinguished by a black band below its neck.

BLACK BAND
It sets this species apart.

DISTRIBUTION OF NAJA SPECIES IN ASIA

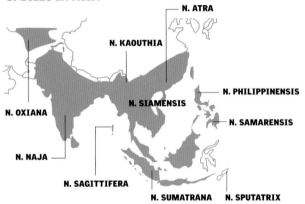

N. ATRA
N. KAOUTHIA
N. SIAMENSIS
N. PHILIPPINENSIS
N. SAMARENSIS
N. OXIANA
N. NAJA
N. SAGITTIFERA
N. SUMATRANA
N. SPUTATRIX

HOW TO DISTINGUISH AMONG THEM

Although the Asian species look similar to one another, they often have distinguishing colors and scale patterns. The simplest way to identify them is by the pattern on their hoods—if you have time to look!

INDIAN COBRA
Naja naja

CHINESE COBRA
Naja atra

ANDAMAN COBRA
Naja sagittifera

GOLDEN SPITTING COBRA
Naja sumatrana

270
SPECIES OF ELAPIDS EXIST WORLDWIDE.

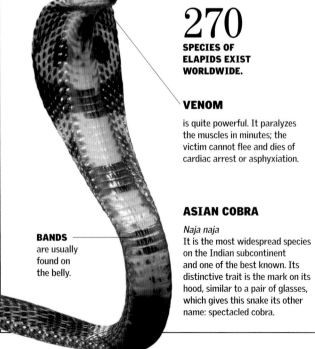

VENOM
is quite powerful. It paralyzes the muscles in minutes; the victim cannot flee and dies of cardiac arrest or asphyxiation.

ASIAN COBRA
Naja naja
It is the most widespread species on the Indian subcontinent and one of the best known. Its distinctive trait is the mark on its hood, similar to a pair of glasses, which gives this snake its other name: spectacled cobra.

BANDS
are usually found on the belly.

The Hood

It is believed that when cobras feel threatened or are about to attack, they spread their hoods by widening their necks in order to look larger than they actually are. The mechanism involves the ribs, which are widened by the muscles that lie between them. When cobras put on this display, they are ready to strike. Some species also hiss while in this position.

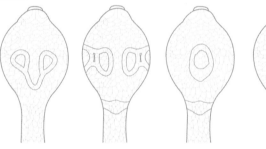

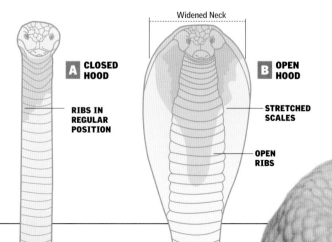

Widened Neck

A CLOSED HOOD

RIBS IN REGULAR POSITION

B OPEN HOOD

STRETCHED SCALES

OPEN RIBS

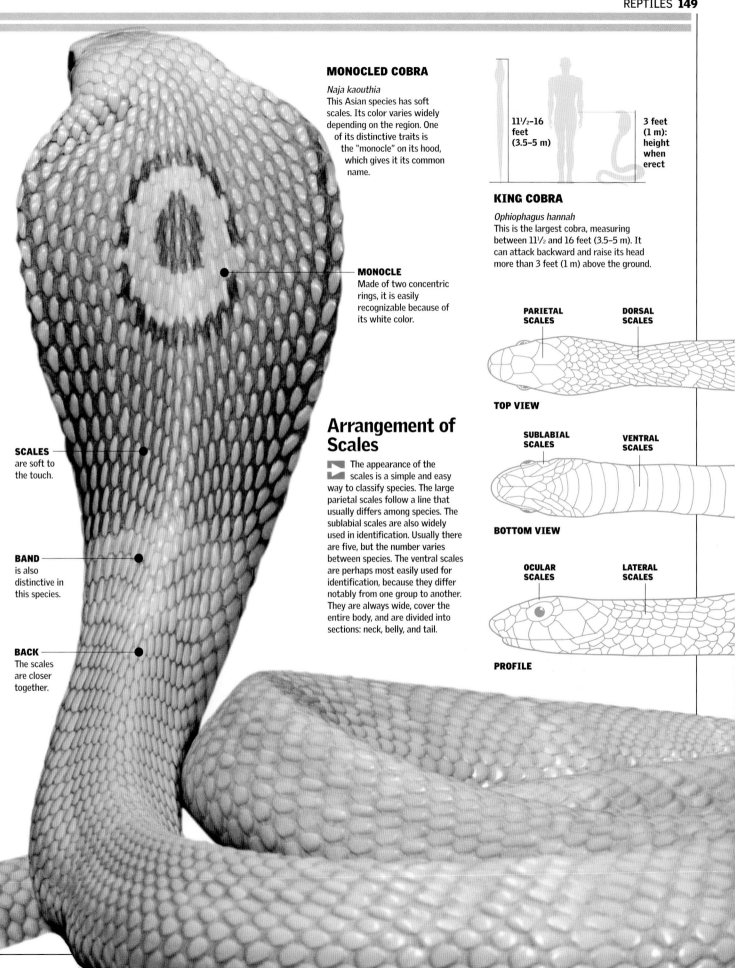

MONOCLED COBRA

Naja kaouthia
This Asian species has soft scales. Its color varies widely depending on the region. One of its distinctive traits is the "monocle" on its hood, which gives it its common name.

MONOCLE
Made of two concentric rings, it is easily recognizable because of its white color.

SCALES
are soft to the touch.

BAND
is also distinctive in this species.

BACK
The scales are closer together.

11½–16 feet (3.5–5 m)

3 feet (1 m): height when erect

KING COBRA

Ophiophagus hannah
This is the largest cobra, measuring between 11½ and 16 feet (3.5–5 m). It can attack backward and raise its head more than 3 feet (1 m) above the ground.

Arrangement of Scales

The appearance of the scales is a simple and easy way to classify species. The large parietal scales follow a line that usually differs among species. The sublabial scales are also widely used in identification. Usually there are five, but the number varies between species. The ventral scales are perhaps most easily used for identification, because they differ notably from one group to another. They are always wide, cover the entire body, and are divided into sections: neck, belly, and tail.

PARIETAL SCALES
DORSAL SCALES
TOP VIEW

SUBLABIAL SCALES
VENTRAL SCALES
BOTTOM VIEW

OCULAR SCALES
LATERAL SCALES
PROFILE

4 FISH AND AMPHIBIANS

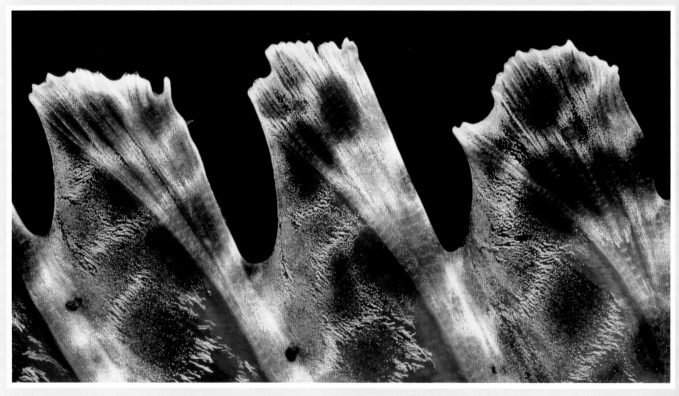

Fish: General Characteristics

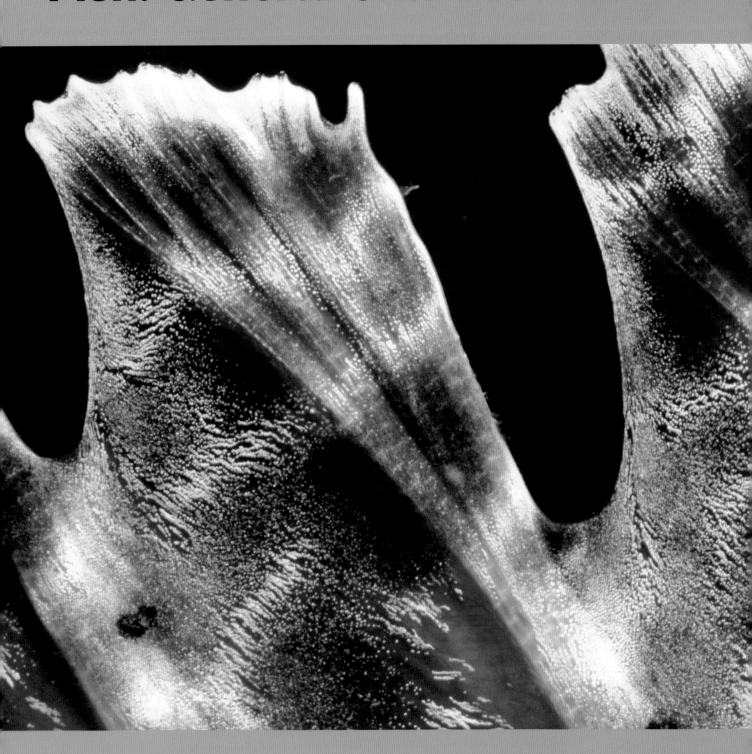

Fish were the first vertebrates with bony skeletons to appear on the earth. They doubtless form the most numerous group of vertebrates.

Unlike today's fish, the earliest fish had no scales, fins, or jawbone, but they did have a type of dorsal fin. Over time, they have been changing in form and size to adapt to different

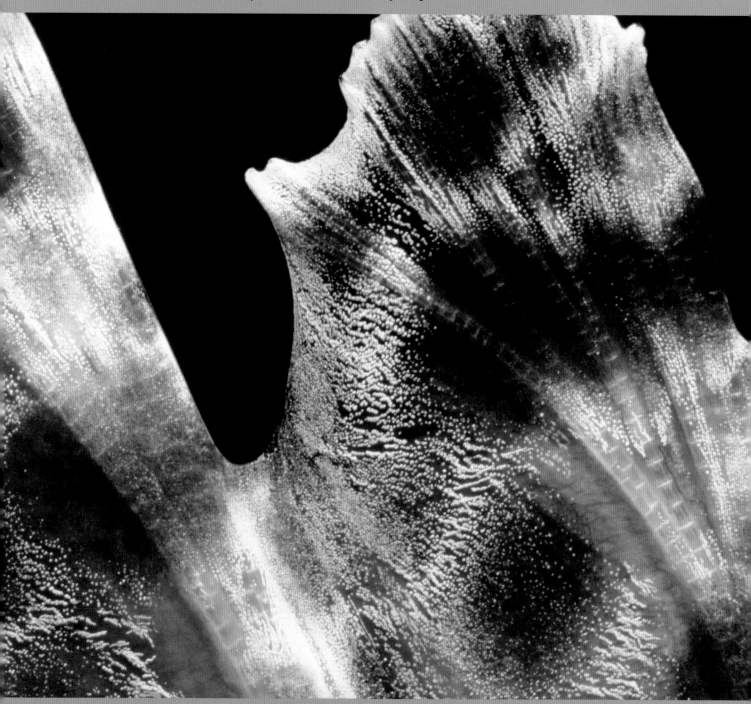

environments, in both fresh water and salt water. Their bodies are generally streamlined, being covered with smooth scales and having fins that enable them to move with energy, direction, and stability. In place of lungs, these complex creatures normally breathe through gills that capture oxygen dissolved in the water, and they are cold-blooded.

Distinguishing Features

Similar characteristics define nearly all fish, with a few rare exceptions. These aquatic animals are designed to live underwater, and they have a jawbone and lidless eyes and are cold-blooded. They breathe through gills and are vertebrates—that is, they have a spinal column. They live in the oceans, from the poles to the equator, as well as in bodies of fresh water and in streams. Some fish migrate, but very few can pass from salt water to fresh water or vice versa. Their fins enable them to swim and move in different directions. Animals such as dolphins, seals, and whales, are at times mistaken for fish, but they are actually mammals.

NASAL PITS
Also called nares; lie on either side of the head

EYES
On the side of the head, protected by fatty membranes

HEAD
One of the three main divisions of its body

PECTORAL FIN
Symmetrical, relatively small, and with a radial structure

ANTERIOR DORSAL FIN
This fin has stiff rays and has a stabilizing function.

MOUTH
The angle of the mouth affects what the fish can eat.

OPERCULUM
A bony flap that covers the gills and helps regulate water flow

GILLS
The fish's breathing organs

PELVIC FINS
These permit the fish to swim upward and downward.

Gill Breathing

Gills are the organs that fish use to breathe. They are made of filaments linked by the gill arches. The fish uses its gills to take in oxygen dissolved in the water. Through a process known as diffusion, oxygen is transferred to the blood, which has a lower concentration of oxygen than the water. In this way the fish oxygenates its blood, which then circulates to the rest of its body. In most bony fish (osteichthyes), water flows in through the mouth, splits into two streams, and exits through the gill slits.

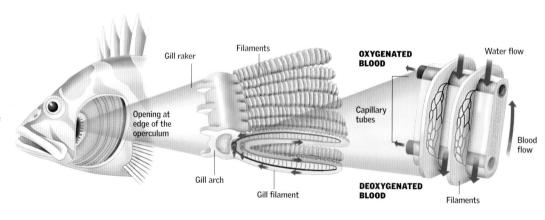

Gill raker

Filaments

Opening at edge of the operculum

Gill arch

Gill filament

OXYGENATED BLOOD

Capillary tubes

Water flow

Blood flow

Filaments

DEOXYGENATED BLOOD

Near fossils

Choanichthyes (Sarcopterygii) are archaic bony fish with fleshy fins. Some of them were the first animals with lungs. Only a few species survive.

COELACANTH

Latimeria chalumnae
This species was thought to have gone extinct millions of years ago, until one was discovered alive off the coast of South Africa in 1938; more of these fish were found later.

Jawless Fish

Of the ancient agnathans, considered the first living vertebrates, only lampreys and hagfish are left.

SEA LAMPREY

Lampetra sp.
Its round, toothed mouth allows it to suck the blood of fish of various species. There are also freshwater lampreys.

Just Cartilage

Cartilaginous fish, such as rays and sharks, have extremely flexible skeletons with little or no bone.

RAY

Raja miraletus
Its large fins send currents of water carrying plankton and small fish to its mouth. The ray is very fast.

SCALES
The scales are imbricate—that is, they overlap one another.

POSTERIOR DORSAL FIN
This soft-structured fin is located between the dorsal fin and the tail.

LATERAL LINE
Fish have sensory organs all along this line.

With Spines

Osteichthyes is the most numerous class of fish. The skeleton has some level of calcification.

ATLANTIC MACKEREL

Scomber scombrus
This fish has no teeth. It lives in temperate waters, and its meat is considered delicious. It can live for more than ten years.

ANAL FIN
Soft, with a row of finlets

TAIL MUSCLE
This is the strongest muscle in the fish.

CAUDAL FIN
It moves from side to side, propelling the fish forward.

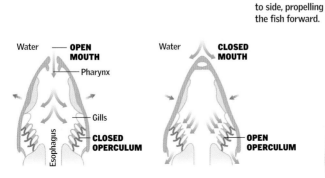

IN ACTION
Water enters the mouth and flows over the gills. After the gills extract oxygen, the water is expelled through the gill slits.

OPERCULUM
Opens and closes the openings where water exits

Water — OPEN MOUTH
Pharynx
Gills
Esophagus
CLOSED OPERCULUM

Water — CLOSED MOUTH
OPEN OPERCULUM

25,000

IS THE NUMBER OF KNOWN FISH SPECIES, MAKING UP NEARLY ONE HALF OF ALL CHORDATE SPECIES.

Bony Fish

The group of fish that have evolved and diversified most in the last few million years are the osteichthyes, fish with spines and jawbones. In general, their skeletons are relatively small but firm, being made mostly of bone. Flexible fins enable them to control their movements with precision. The various species of osteichthyes have adapted to a wide variety of environments and even to extreme conditions.

Solid Structure

The skeleton of a bony fish is divided into the cranium, spinal column, and fins. The opercula, which cover their gills, are also made of bone. The cranium holds the brain and supports the jawbone and gill arches. The vertebrae of the spine are jointed; they provide support to the body and join the ribs at the abdomen.

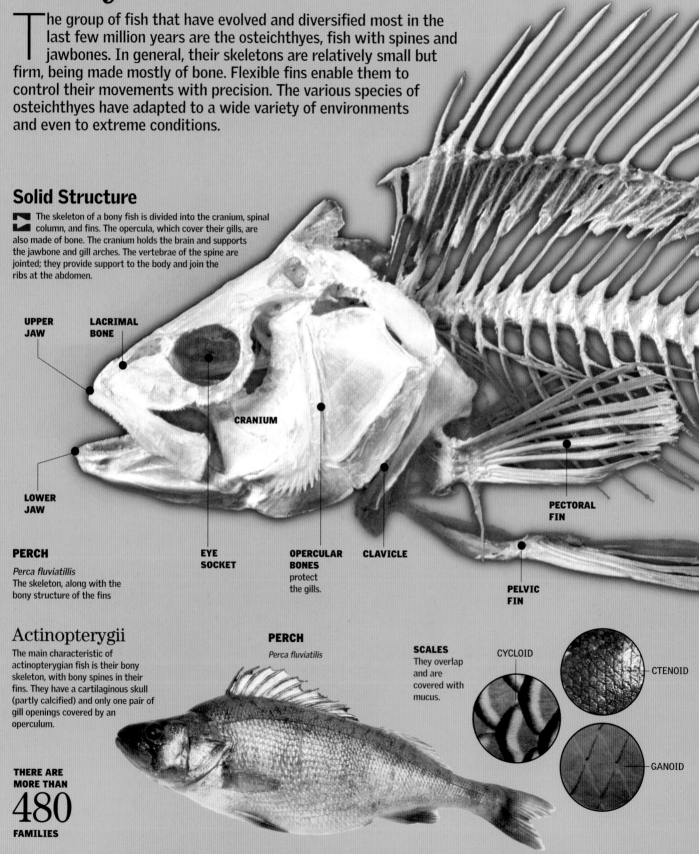

UPPER JAW

LACRIMAL BONE

CRANIUM

LOWER JAW

EYE SOCKET

OPERCULAR BONES
protect the gills.

CLAVICLE

PECTORAL FIN

PELVIC FIN

PERCH

Perca fluviatillis
The skeleton, along with the bony structure of the fins

Actinopterygii

The main characteristic of actinopterygian fish is their bony skeleton, with bony spines in their fins. They have a cartilaginous skull (partly calcified) and only one pair of gill openings covered by an operculum.

PERCH
Perca fluviatilis

SCALES
They overlap and are covered with mucus.

CYCLOID

CTENOID

GANOID

THERE ARE MORE THAN

480

FAMILIES

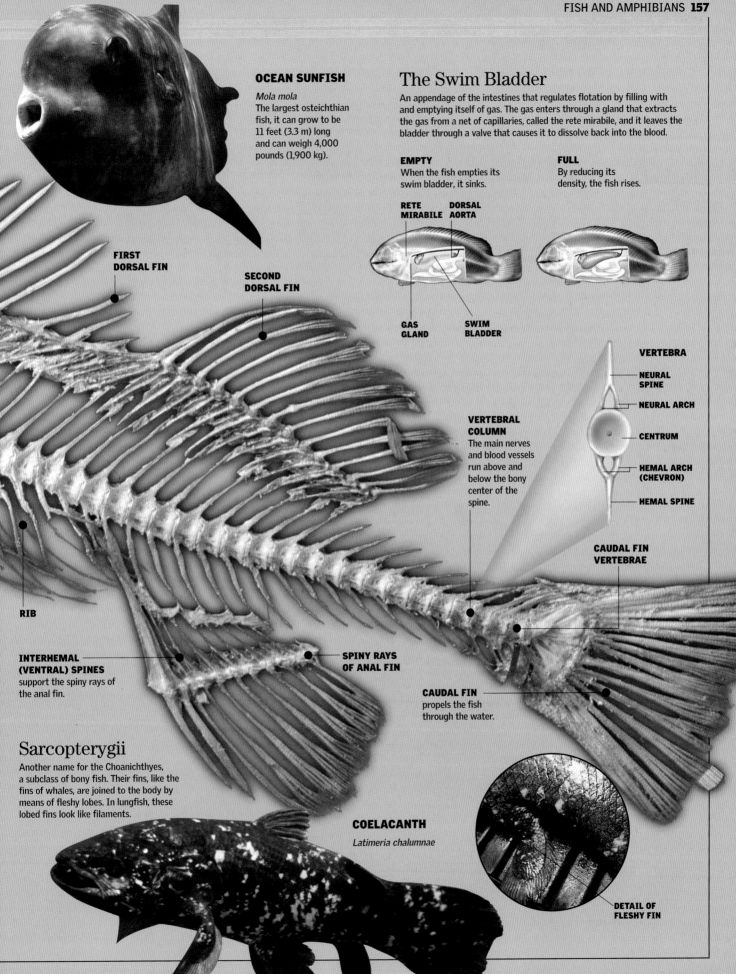

OCEAN SUNFISH

Mola mola
The largest osteichthian fish, it can grow to be 11 feet (3.3 m) long and can weigh 4,000 pounds (1,900 kg).

The Swim Bladder

An appendage of the intestines that regulates flotation by filling with and emptying itself of gas. The gas enters through a gland that extracts the gas from a net of capillaries, called the rete mirabile, and it leaves the bladder through a valve that causes it to dissolve back into the blood.

EMPTY
When the fish empties its swim bladder, it sinks.

FULL
By reducing its density, the fish rises.

RETE MIRABILE
DORSAL AORTA
GAS GLAND
SWIM BLADDER

FIRST DORSAL FIN

SECOND DORSAL FIN

VERTEBRA
NEURAL SPINE
NEURAL ARCH
CENTRUM
HEMAL ARCH (CHEVRON)
HEMAL SPINE

VERTEBRAL COLUMN
The main nerves and blood vessels run above and below the bony center of the spine.

CAUDAL FIN VERTEBRAE

RIB

INTERHEMAL (VENTRAL) SPINES
support the spiny rays of the anal fin.

SPINY RAYS OF ANAL FIN

CAUDAL FIN
propels the fish through the water.

Sarcopterygii

Another name for the Choanichthyes, a subclass of bony fish. Their fins, like the fins of whales, are joined to the body by means of fleshy lobes. In lungfish, these lobed fins look like filaments.

COELACANTH

Latimeria chalumnae

DETAIL OF FLESHY FIN

Cartilaginous Fish

As indicated by the name, the skeleton of cartilaginous fish is made of cartilage, a flexible, durable substance that is softer than bone. They have jaws and teeth, which are usually hard and sharp. Their body is covered with hard scales. However, they lack a characteristic shared by most bony fish—the swim bladder, an organ that helps fish to float. Their pectoral fins, tail, and flat head give this group a streamlined profile.

Sharks

These fish live in tropical waters, although some do inhabit temperate waters or fresh water. They have an elongated, cylindrical shape and a pointed snout, with the mouth on the underside. Each side of their head has five to seven gill slits.

2,650 pounds
(1.2 metric tons)
NORMAL WEIGHT OF A SHARK
(SUPERORDER SELACHIMORPHA)

LIGHT AND FLEXIBLE
The skeleton is very flexible, but the spinal column of cartilage is firm, with mineral deposits.

SPINAL COLUMN

NOSTRIL

BLOOD
They are cold-blooded.

SHARP TEETH
The teeth are triangular in shape. All chondrichthyes lose their teeth and grow new ones.

SURFACE PORE

EPIDERMIS

HEAT-GENERATING MUSCLES

SENSORY CELLS

NERVES

GELATINOUS TRACT

ACUTE SENSES
Chondrichthyes have ampullae of Lorenzini, acutely sensitive lateral lines, and a highly developed sense of smell.

AMPULLAE OF LORENZINI
detect electric signals transmitted by potential prey.

Primitive

The ancient origin of Chondrichthyes contrasts sharply with their highly evolved senses. This is a fossilized cartilage vertebra of a shark from the Paleozoic era, between 245 and 540 million years ago. It was found in a fossil deposit in Kent, England. The blood of sharks has a high concentration of urea, which is presumed to be an adaptation to salt water and constitutes a fundamental difference between sharks and their freshwater ancestors.

Manta Rays and Skates

These fish have two pectoral fins joined on the front of the body. They use them to swim, giving the impression that they fly in the water. The rest of the body moves similarly to a whip. Their eyes are located on the upper side of the body; the mouth and gills are on the lower side.

RAY

Raja clavata (Thornback Ray)
This species lives in cold oceans in depths up to 660 feet (200 m).

Rays may have five or six rows of gills; chimaeras have only one.

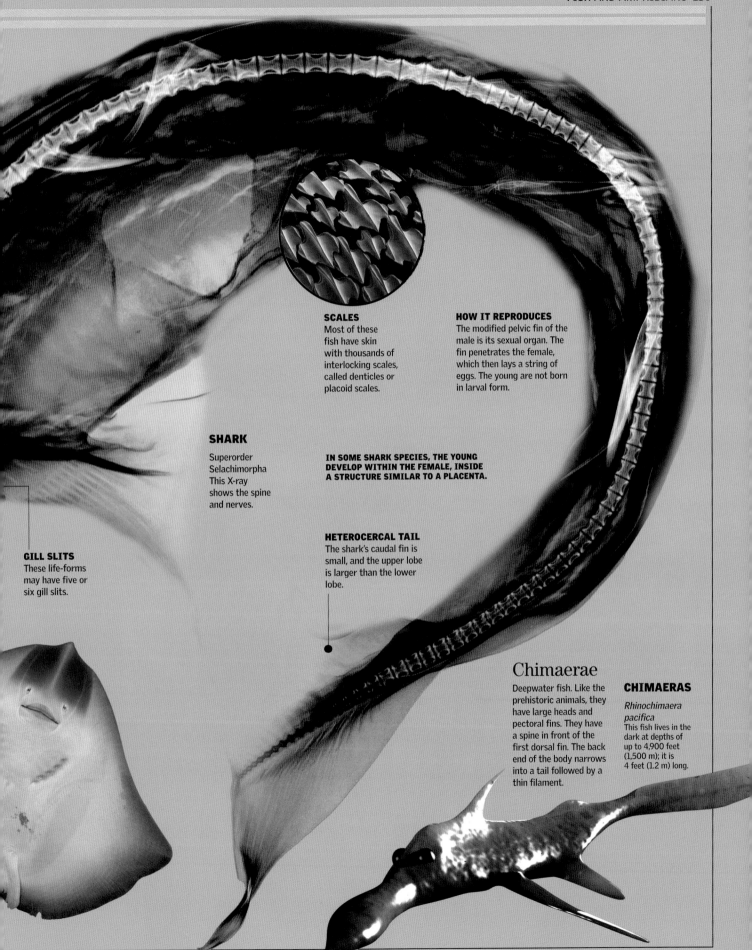

SCALES
Most of these fish have skin with thousands of interlocking scales, called denticles or placoid scales.

HOW IT REPRODUCES
The modified pelvic fin of the male is its sexual organ. The fin penetrates the female, which then lays a string of eggs. The young are not born in larval form.

SHARK
Superorder Selachimorpha This X-ray shows the spine and nerves.

IN SOME SHARK SPECIES, THE YOUNG DEVELOP WITHIN THE FEMALE, INSIDE A STRUCTURE SIMILAR TO A PLACENTA.

HETEROCERCAL TAIL
The shark's caudal fin is small, and the upper lobe is larger than the lower lobe.

GILL SLITS
These life-forms may have five or six gill slits.

Chimaerae
Deepwater fish. Like the prehistoric animals, they have large heads and pectoral fins. They have a spine in front of the first dorsal fin. The back end of the body narrows into a tail followed by a thin filament.

CHIMAERAS

Rhinochimaera pacifica
This fish lives in the dark at depths of up to 4,900 feet (1,500 m); it is 4 feet (1.2 m) long.

Anatomy

Most fish have the same internal organs as amphibians, reptiles, birds, and mammals. The skeleton acts as a support, and the brain receives information through the eyes and the lateral line to coordinate the motions of the muscles in propelling the fish through the water. Fish breathe with gills, they have a digestive system designed to transform food into nutrients, and they have a heart that pumps blood through a network of blood vessels.

SIMPLE EYE
Each eye focuses to one side; there is no binocular vision.

Suspensory ligament — Retina
LENS — **OPTIC NERVE**
Iris

BRAIN
receives information and coordinates all the fish's actions and functions.

Cyclostomata

Its digestive tract is little more than a straight tube extending from its round, jawless mouth to the anus. Because of their simplicity, many species of lampreys are parasites. They live off the blood of other fish and have thin pharyngeal sacs instead of gills.

45 **THE CURRENT NUMBER OF SPECIES OF CYCLOSTOMATA**

MOUTH

GILLS
Structures with multiple folds that provide oxygen to the blood

HEART
receives all the blood and pumps it toward the gills.

LIVER

CAUDAL FIN

ANUS

LAMPREY
Lampetra sp.

BREATHING SACS

EYE

HEART

LIVER

TOOTHED MOUTH

FIRST DORSAL FIN

INTESTINE

SUPPORT FOR PHARYNGEAL SACS

NOTOCHORD

TESTICLES

VERTEBRAE

BRAIN

RIGHT KIDNEY

STOMACH

GONAD

Chondrichthyes

A shark has the same organic structures as a bony fish, except for the swim bladder. A shark also has a corkscrewlike structure called a spiral valve at the end of its intestine to increase the surface area for absorption of nutrients.

NASAL PIT

SHARK
Carcharodon sp.

MOUTH

GILL SLITS

HEART

LIVER

STOMACH

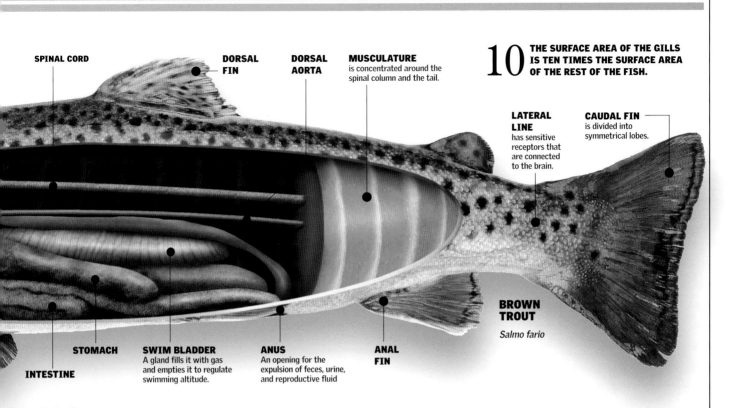

SPINAL CORD

DORSAL FIN

DORSAL AORTA

MUSCULATURE
is concentrated around the spinal column and the tail.

10 THE SURFACE AREA OF THE GILLS IS TEN TIMES THE SURFACE AREA OF THE REST OF THE FISH.

LATERAL LINE
has sensitive receptors that are connected to the brain.

CAUDAL FIN
is divided into symmetrical lobes.

BROWN TROUT
Salmo fario

STOMACH

INTESTINE

SWIM BLADDER
A gland fills it with gas and empties it to regulate swimming altitude.

ANUS
An opening for the expulsion of feces, urine, and reproductive fluid

ANAL FIN

Osteichthyes

Typically, their organs are compressed in the lower front quarter of the body. The rest of their internal structure consists mainly of the muscles that the fish uses to swim. Some bony fish, such as carp, have no stomach but rather a tightly coiled intestine.

REGULATION OF SALINITY

FRESHWATER FISH
Freshwater fish run the risk of losing salt to their environment. They drink only a small quantity of water, and they obtain additional salt from their food.

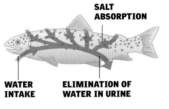

SALT ABSORPTION

WATER INTAKE

ELIMINATION OF WATER IN URINE

SALTWATER FISH
These fish constantly absorb salt water to replenish the water in their bodies, but they must eliminate excess salt from the marine environment.

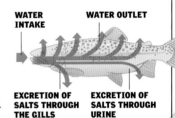

WATER INTAKE

WATER OUTLET

EXCRETION OF SALTS THROUGH THE GILLS

EXCRETION OF SALTS THROUGH URINE

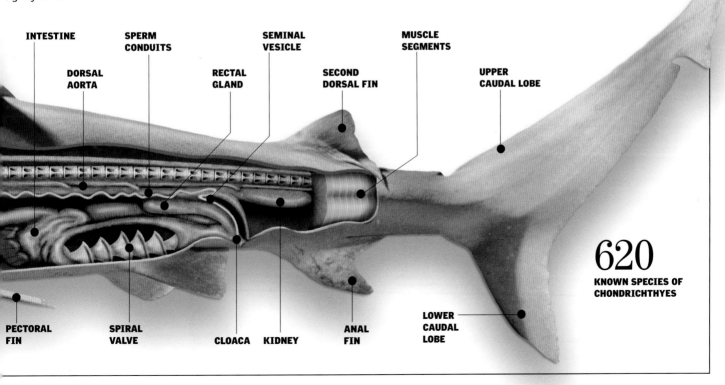

INTESTINE

DORSAL AORTA

SPERM CONDUITS

RECTAL GLAND

SEMINAL VESICLE

SECOND DORSAL FIN

MUSCLE SEGMENTS

UPPER CAUDAL LOBE

620

KNOWN SPECIES OF CHONDRICHTHYES

PECTORAL FIN

SPIRAL VALVE

CLOACA

KIDNEY

ANAL FIN

LOWER CAUDAL LOBE

Protective Layer

Most fish are covered with scales, an external layer of transparent plates. All fish of a given species have the same number of scales. Depending on the family and genus of a fish, its scales can have a variety of characteristics. Scales on the lateral line of the body have small orifices that link the surface with a series of sensory cells and nerve endings. It is also possible to determine a fish's age by studying its scales.

FOSSILIZED SCALES

The remains of these thick, shiny, enameled scales belong to the extinct genus *Lepidotes*, a fish that lived during the Mesozoic era.

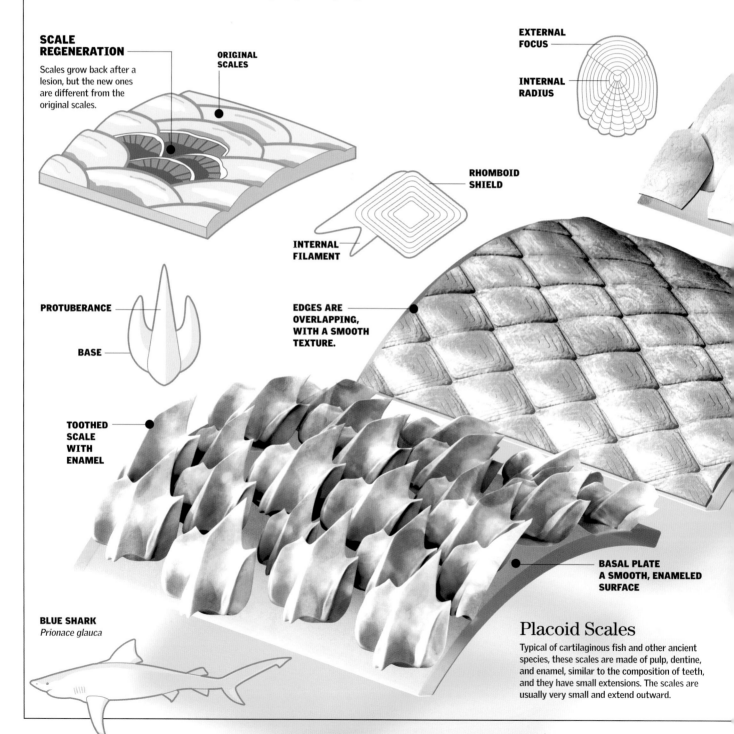

SCALE REGENERATION

Scales grow back after a lesion, but the new ones are different from the original scales.

ORIGINAL SCALES

EXTERNAL FOCUS

INTERNAL RADIUS

RHOMBOID SHIELD

INTERNAL FILAMENT

PROTUBERANCE

BASE

EDGES ARE OVERLAPPING, WITH A SMOOTH TEXTURE.

TOOTHED SCALE WITH ENAMEL

BASAL PLATE A SMOOTH, ENAMELED SURFACE

BLUE SHARK
Prionace glauca

Placoid Scales

Typical of cartilaginous fish and other ancient species, these scales are made of pulp, dentine, and enamel, similar to the composition of teeth, and they have small extensions. The scales are usually very small and extend outward.

FOCUS

TOOTHED
SPOKES

EPIDERMIS
WITH PROTECTIVE
MUCUS

EPIDERMIS
COVERS MOST
OF THE BODY.

TOOTHED EDGES
PROVIDE
ROUGHNESS.

Ctenoid Scales

These scales overlap like tiles on a roof,
the same as cycloid scales. Another
very common type of scale among
bony fish, they are rough, having small
extensions that look like combs.

SHIELDS
THE STURGEON HAS
FIVE ROWS OF THESE.

PERCH
Perca sp.

Cycloid Scales

One of the most common types of scales
among bony fish, the cycloid scales are
organized so that the exposed surfaces
overlap, forming a smooth and flexible cover.
They are round with a soft, exposed surface,
such as those of carps and silversides.

AGE BY SCALES

A fish does not add new scales as it grows,
but the scales it has increase in size. In
this way, growth rings are formed, and the
rings reveal the age of the specimen.

CUTICLE
HAS A MUCOUS
CONSISTENCY.

SALMON
Family Salmonidae

WINTER
GROWTH
LINE

SUMMER
GROWTH
LINE

Ganoid Scales

Rhomboid in shape, these scales are interwoven
and connected with fibers. The name comes from
their outer covering, which is a layer of ganoin, a
type of shiny enamel. Sturgeon and pipefish have
scales of this type.

DISTRIBUTION OF SCALES

Most scales occur in rows that slant diagonally
downward and back. Species can be accurately
identified by the number of rows (as counted
along the lateral line), among other characteristics.

EXPOSED
AREA

STURGEON
Acipenser sturio

TRANSVERSE
LINE

LATERAL
LINE

RED SNAPPER
Lutjanus campechanus

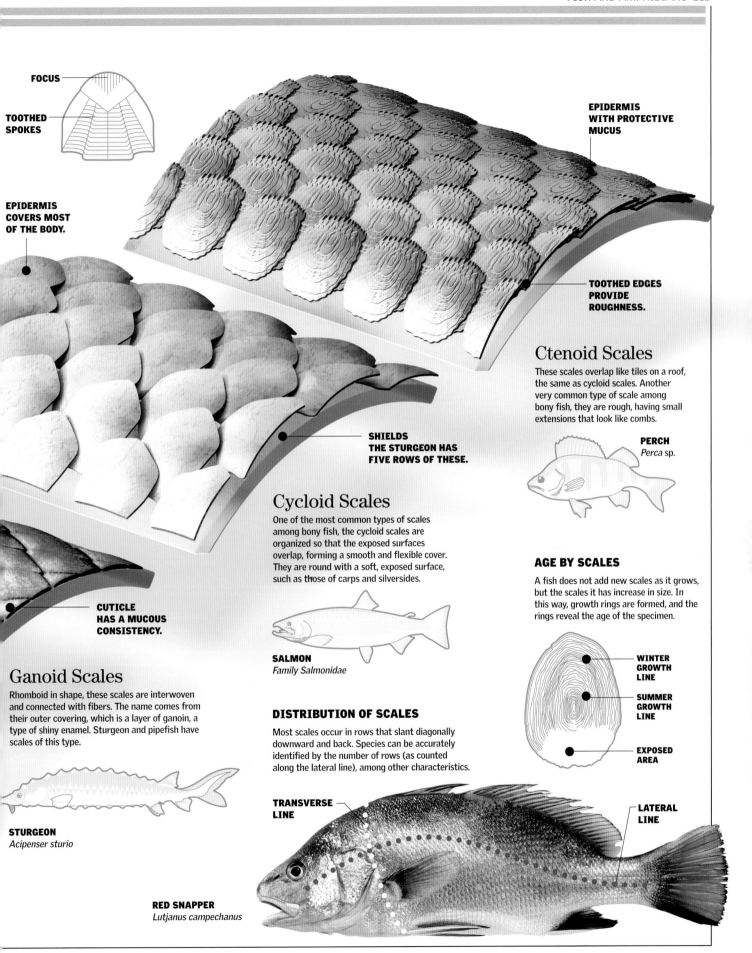

The Art of Swimming

To swim, fish move in three dimensions: forward and back, left and right, and up and down. The main control surfaces that fish use for maneuvering are the fins, including the tail, or caudal fin. To change direction, the fish tilts the control surfaces at an angle to the water current. The fish must also keep its balance in the water; it accomplishes this by moving its paired and unpaired fins.

UPSIDE-DOWN CATFISH

Synodontis nigriventris
This fish swims upside down, seeking food sources that are less accessible to other species.

MUSCLES

The tail has powerful muscles that enable it to move like an oar.

GREAT WHITE SHARK
Carcharodon carcharias

RED MUSCLES
are for slow or regular movements.

LARGER WHITE
muscles are for moving with speed, but they tire easily.

1 Starting Out

The movement of a fish through the water is like that of a slithering snake. Its body goes through a series of wavelike movements similar to an S curve. This process begins when the fish moves its head slightly from side to side.

The crest of the body's wave moves from back to front.

In its side-to-side movement, the tail displaces the water.

At first, the tail is even with the head.

Streamlined Shape

Like the keel of a ship, the rounded contours of a fish are instrumental. In addition, most of a fish's volume is in the front part of its body. As the fish swims forward, its shape causes the density of the water ahead to be reduced relative to the density of the water behind. This reduces the water's resistance.

The head moves from side to side.

THE FISH'S KEEL

A ship has a heavy keel in the lower part to keep it from capsizing. Fish, on the other hand, have the keel on top. If the paired fins stop functioning to keep the fish balanced, the fish turns over because its heaviest part tends to sink, which happens when fish die.

KEEL

LIVE FISH   **DEAD FISH**

THE FASTEST

SAILFISH
Istiophorus platypterus

The powerful caudal fin displaces large amounts of water.

The unfurled dorsal fin can be up to 150 percent of the width of the fish's body.

Its long upper jaw enables it to slice through the water, aiding this fish's hydrodynamics.

70 miles per hour
(109 km/h) THE MAXIMUM SWIMMING SPEED IT ATTAINS

Forward Motion

results from the synchronized S-curve movement of the muscles surrounding the spinal column. These muscles usually make alternating lateral motions. Fish with large pectoral fins use them like oars for propulsion.

The oarlike movement of the tail is the main force used for forward motion.

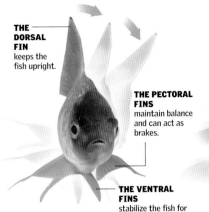

THE DORSAL FIN keeps the fish upright.

THE PECTORAL FINS maintain balance and can act as brakes.

THE VENTRAL FINS stabilize the fish for proper balance.

Balance

When the fish is moving slowly or is still in the water, the fins can be seen making small movements to keep the body in balance.

Upward and Downward

The angle of the fins relative to the body allows the fish to move up or down. The paired fins, located in front of the center of gravity, are used for this upward or downward movement.

ASCENT

PAIRED FINS

DESCENT

② Forceful Stroke

Muscles on both sides of the spinal column, especially the tail muscles, contract in an alternating pattern. These contractions power the wavelike movement that propels the fish forward. The crest of the wave reaches the pelvic and dorsal fins.

The crest of the wave passes to the first dorsal fins.

When the crest reaches the area between the two dorsal fins, the tail fin begins its push to the right.

③ Complete Cycle

When the tail moves back toward the other side and reaches the far right, the head will once again turn to the right to begin a new cycle.

1 second
THE AMOUNT OF TIME IT TAKES FOR THIS SHARK TO COMPLETE ONE SWIMMING CYCLE

CAT SHARK
Scyliorhinus sp.

The resulting impulse moves the fish forward.

Swimming in Groups

Only bony fish can swim in highly coordinated groups. Schools of fish include thousands of individuals that move harmoniously as if they were a single fish. To coordinate their motion, they use their sight, hearing, and lateral line senses. Swimming in groups has its advantages: it is harder to be caught by a predator, and it is easier to find companions or food.

SCHOOL
A group of fish, usually of the same species, that swim together in a coordinated manner and with specific individual roles

1 cubic mile
(4 cu km)
THE AREA THAT CAN BE TAKEN UP BY A SCHOOL OF HERRING

The fish on the outside, guided by those in the middle, are in charge of keeping the group safe.

The fish in the middle control the school.

Life Cycle

In an underwater environment, animals can simply secrete their sex cells into the water. But for fertilization to be effective, the male and the female must synchronize their activities. Many species, such as the salmon, travel great distances to meet with potential mates. Upon meeting a mate they release their sex cells. The time and place are important because the survival of the eggs depends on the water temperature. Parent-child relations are extremely varied, from complete neglect of the eggs once laid to constant watchfulness and protection of the young.

External Fertilization

In most fish, fertilization is external to the female's body. The male secretes sperm onto the eggs as soon as they leave the female's body. Typically, the young hatch from the eggs as larvae. Salmon is one species that reproduces this way.

MALE SALMON

FEMALE SALMON

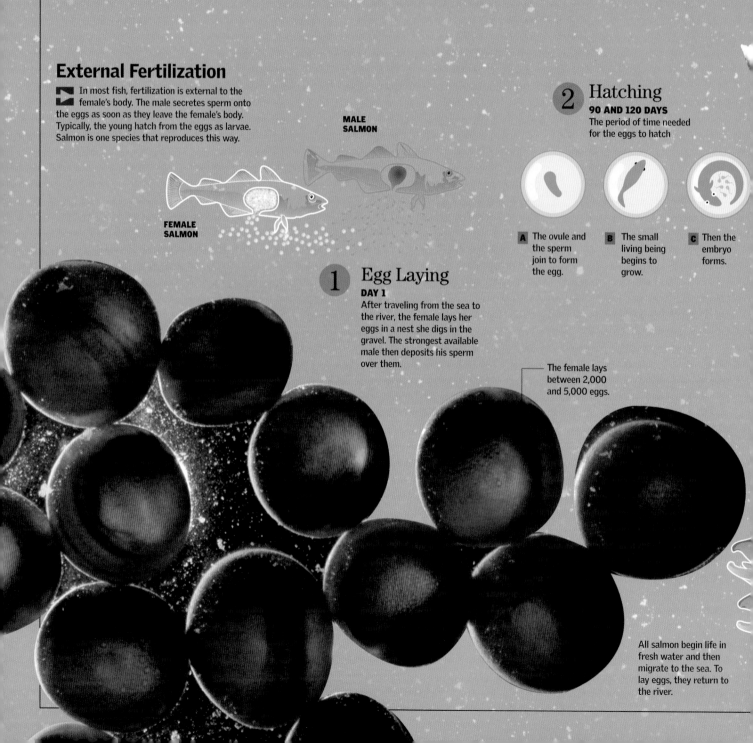

2 Hatching
90 AND 120 DAYS
The period of time needed for the eggs to hatch

A The ovule and the sperm join to form the egg.

B The small living being begins to grow.

C Then the embryo forms.

1 Egg Laying
DAY 1
After traveling from the sea to the river, the female lays her eggs in a nest she digs in the gravel. The strongest available male then deposits his sperm over them.

The female lays between 2,000 and 5,000 eggs.

All salmon begin life in fresh water and then migrate to the sea. To lay eggs, they return to the river.

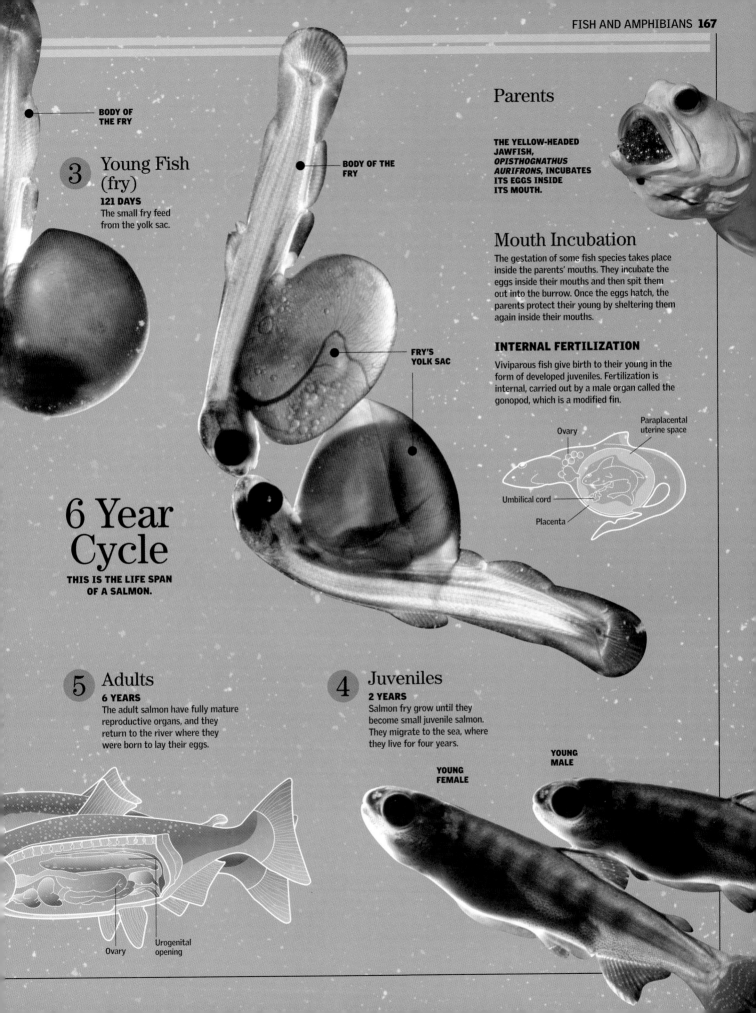

**BODY OF
THE FRY**

3 Young Fish
(fry)
121 DAYS
The small fry feed
from the yolk sac.

BODY OF THE
FRY

FRY'S
YOLK SAC

6 Year
Cycle

**THIS IS THE LIFE SPAN
OF A SALMON.**

Parents

**THE YELLOW-HEADED
JAWFISH,
*OPISTHOGNATHUS
AURIFRONS*, INCUBATES
ITS EGGS INSIDE
ITS MOUTH.**

Mouth Incubation

The gestation of some fish species takes place
inside the parents' mouths. They incubate the
eggs inside their mouths and then spit them
out into the burrow. Once the eggs hatch, the
parents protect their young by sheltering them
again inside their mouths.

INTERNAL FERTILIZATION

Viviparous fish give birth to their young in the
form of developed juveniles. Fertilization is
internal, carried out by a male organ called the
gonopod, which is a modified fin.

Ovary

Paraplacental
uterine space

Umbilical cord

Placenta

5 Adults
6 YEARS
The adult salmon have fully mature
reproductive organs, and they
return to the river where they
were born to lay their eggs.

4 Juveniles
2 YEARS
Salmon fry grow until they
become small juvenile salmon.
They migrate to the sea, where
they live for four years.

YOUNG
MALE

YOUNG
FEMALE

Ovary

Urogenital
opening

Matters of Life and Death

To survive, most fish need adaptations to enable them to flee from their predators or to find food. The European plaice can lie on the ocean floor with its flat body. Its ivory color makes it almost invisible. The flying fish, on the other hand, developed pectoral fins to lift itself up over the surface of the water and flee its enemies.

European Plaice

The European plaice (*Pleuronectes platessa*) is a flat fish with a shape especially designed to allow it to remain motionless on the seafloor. It also provides an example of mimesis. Its two sides are very different. The top side is pigmented with small red spots that camouflage the fish on the seafloor, where it uses its fins to cover itself with sand to hide from predators.

VENTRAL SIDE
remains an ivory color, devoid of pigmentation. This side rests on the ocean floor.

MOUTH
The European plaice's entire body undergoes metamorphosis from its larval state to adulthood. The mouth, however, remains the same.

EUROPEAN PLAICE
Pleuronectes platessa

EYES
Both are located on the right side.

GILLS
The European plaice breathes through its gills.

OPERCULUM
is the bone that supports the gill structure.

Transformation

At birth, the European plaice does not have a flat form but looks like a normal fish. It eats near the surface and swims using its swim bladder. As time goes by, its body becomes flat. The swim bladder dries up, and the fish sinks to the bottom of the sea.

1 **5 days**
⅛ inch (3.5 mm)

The vertebrae begin to form.

One eye on each side

2 **10 days**
⅛ inch (4 mm)

The fold of the fin is forming, and the mouth is already open.

3 **22 days**
⅓ inch (8 mm)

The cleft of the tail develops.

The left eye moves to the top of the head.

45 days

IS THE AMOUNT OF TIME THE EUROPEAN PLAICE TAKES TO BECOME A FLAT FISH FROM A TYPICAL STREAMLINED LARVA.

Flying Fish

Exocoetidae, or flying fish, are a family of ocean fish that includes 52 species grouped in eight genera. They are found in all the oceans, especially in warm tropical and subtropical waters. Their most surprising characteristic is their unusually large pectoral fins, which give them the ability to fly and glide for short distances.

1

ESCAPE
When a predator appears, the flying fish propels itself out of the water.

2

TAKEOFF
The fish comes to the surface and elevates itself as high as it can, skipping over the water.

They reach heights of up to 19 feet (6 m).

3

GLIDING
The average gliding distance is 160 feet (50 m), but they can glide as far as 660 feet (200 m).

These fish cover distances of up to 160 feet (50 m) in the air.

Flying fish measure from 7 to 18 inches (18 to 45 cm) long.

ANATOMY
This fish slides over the water with its hardened fins, and it can reach speeds up to 40 miles per hour (65 km/h) for as long as 30 seconds.

This fish has highly developed pectoral and pelvic fins.

FLYING FISH
Exocoetus volitans

SPOTS
are useful for camouflage in the sand and for hiding from its predators.

Scorpion Fish

Found in the reefs of the Gulf of Mexico, *Scorpaena plumieri*, known commonly as the scorpion fish, has a brown, spotted body with many appendages that look like moss between its mouth and its eyes. This fish is hard to see because its texture and color help it blend easily into the seafloor. Its dorsal fins have a powerful venom, which causes intense pain.

FIN
The dorsal, anal, and caudal fins form a continuous line around the body.

CAUDAL FIN
Thin, barely used for swimming.

SCORPION FISH
Scorpaena plumieri

4 **45 days**
½ inch (11 mm)

The pigment cells join to form dark spots.

It no longer looks to the right, but upward.

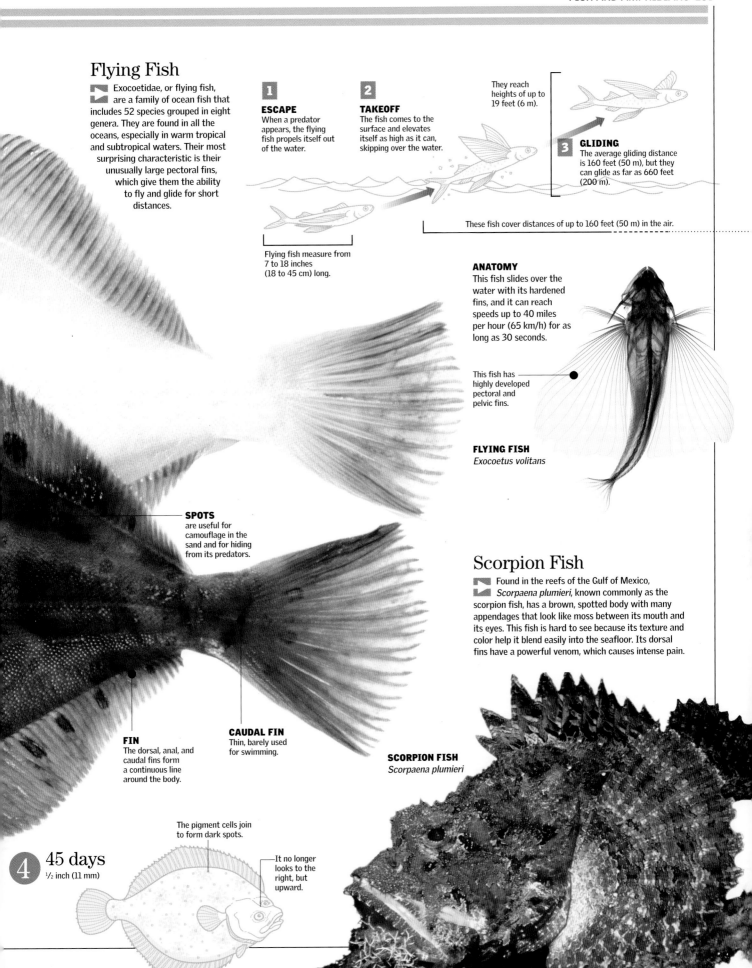

Long and Flexible

The sea horse is a small ocean fish that belongs to the same family as pipefish and sea dragons (Syngnathidae). Its name comes from its horselike head. In fact, no other fish genus has its head at a right angle to the rest of its body. Because it cannot use speed to escape from its predators, the sea horse has the ability to change color to blend in with its environment. The reproduction process of these fish is also very unique. The male has an incubating pouch in which the female deposits the fertilized eggs.

BLACK-STRIPED PIPEFISH
Syngnathus abaster
One of the slowest fish in the sea, the black-striped pipefish moves by means of slight undulations of its pectoral fins, which can vibrate up to 35 times per second.

EYES
Large, for acute vision

NOSE
Pipe-shaped, giving the head a horselike shape

Movement

The body of a sea horse is crammed into an armor of large, rectangular bony plates. They swim very differently than other fish. Adopting an upright position, they use their dorsal fin for propulsion. They do not have an anal fin, but rather a long tail that rolls into a spiral. They use it to hold onto underwater plants.

Classification

Thirty-two species of sea horse have been identified worldwide. Classifying them is at times complicated because individuals of the same species can change color and develop long filaments of skin. The size of adult sea horses varies enormously, from the tiny *Hippocampus minotaur*—a species discovered in Australia that never grows beyond 3/4 inch (1.8 cm) long—to the enormous *Hippocampus ingens*, a species in the Pacific that reaches over 12 inches (30 cm) long. It has no pelvic or caudal fins, but it does have a tiny anal fin.

HEAD

ROLLED UP
The tail rolls up into a curl.

UNROLLED
The tail straightens out by unrolling.

TRUNK
The body is supported by the spinal column.

GRASPING TAIL
With their long tails, sea horses can cling to plants on the seafloor.

TAIL
Can be extended to a fully vertical position

WEEDY SEA DRAGON
Phyllopteryx taeniolatus
Its shape is typical of this family, although its tail is not suitable for grasping, like those of sea horses are, and it has a more elongated profile. Its body is covered with seaweed.

SEAWEED
The fish lets it stick to its body so that it can escape detection.

Camouflage

Because they cannot use speed to escape from predators, sea horses and dragon fish use camouflage as a defense strategy. They change color to blend in with their environment, grow skin filaments shaped like seaweed, and use their heads to climb along the seaweed in which they live, swinging from one plant to another.

LINED SEA HORSE
Hippocampus erectus

HABITAT	Caribbean, Indo-Pacific Ocean
NUMBER OF SPECIES	35
SIZE	7–12 inches (18–30 cm)

GILLS
Sea horses breathe through gills.

PECTORAL FIN
One on each side, for lateral movement

1/3 inch (1 cm)
THE SIZE OF A SEA HORSE AT BIRTH

BONY PLATES
Its body is covered with concentric rings of bone.

DORSAL FIN
Sea horses swim upright, propelled by their dorsal fin.

35 species
OF SEA HORSES LIVE IN THE CARIBBEAN, THE PACIFIC OCEAN, AND THE INDIAN OCEAN.

Reproduction
The male has an incubating pouch in which the female deposits her eggs. The sac closes, and the embryos develop, nourished by the male. He later expels the young, now mature and independent, through a series of contractions.

1 During the mating season, the female lays some 200 eggs in the male's pouch using her egg-depositing organ. There, the eggs are fertilized. When the time for birth arrives, the male clings to seaweed with his tail.

2 The male bends his body backward and forward, as if having contractions. The sac's opening widens, and the birthing process begins. Soon the young begin to appear.

3 As the male's belly contracts, the young sea horses are gradually born. Each one is 1/3 inch (1 cm) long. They begin to feed on phytoplankton right away. The birthing process can last two days, after which the male is exhausted.

Deadly Weapon

One of the greatest predators in the ocean is the great white shark, easily identified by its distinctive white coloring, black eyes, and fierce teeth and jaws. Many biologists believe that attacks on humans result from the shark's exploratory behavior, because these fish often lift their heads above the water and explore things by biting them. This activity is often dangerous because of the sharpness of the sharks' teeth and the strength of their jaws. Great white sharks are implicated in most fatal shark attacks on humans, especially on surfers and divers.

NASAL PITS

EYES
They have poor vision and use their sense of smell to hunt.

JAW
During an attack, it stretches forward.

Senses

Sharks have senses that most animals lack. The ampullae of Lorenzini are small clefts in the shark's head that detect electricity. This sense helps them find prey hidden in the sand. The lateral line is used to detect movement or sound underwater. Smell is their most advanced sense, and it occupies two-thirds of their brain. They also have a highly developed sense of hearing, which allows them to detect very low-frequency sounds.

SHARK ATTACKS 1876–2004

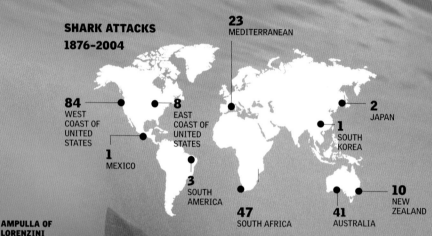

23 MEDITERRANEAN

84 WEST COAST OF UNITED STATES

8 EAST COAST OF UNITED STATES

2 JAPAN

1 SOUTH KOREA

1 MEXICO

3 SOUTH AMERICA

47 SOUTH AFRICA

41 AUSTRALIA

10 NEW ZEALAND

220 ATTACKS IN 128 YEARS

HEARING
Detects sounds of very low frequency

AMPULLA OF LORENZINI
Detects nerve impulses

NOSE
The most highly developed sense is smell; it takes up two thirds of the brain.

LATERAL LINE
detects movements or sounds underwater.

ELECTRIC RADAR

GREAT WHITE SHARK
Carcharodon carcharias

HABITAT	Oceans
WEIGHT	4,400 pounds (2,000 kg)
LENGTH	23 feet (7 m)
LIFE SPAN	30–40 years

CAUDAL FIN
The great white shark has a large heterocercal caudal fin.

PECTORAL FIN
Highly developed and very important for swimming

DORSAL FIN

ANAL FIN

PELVIC FIN

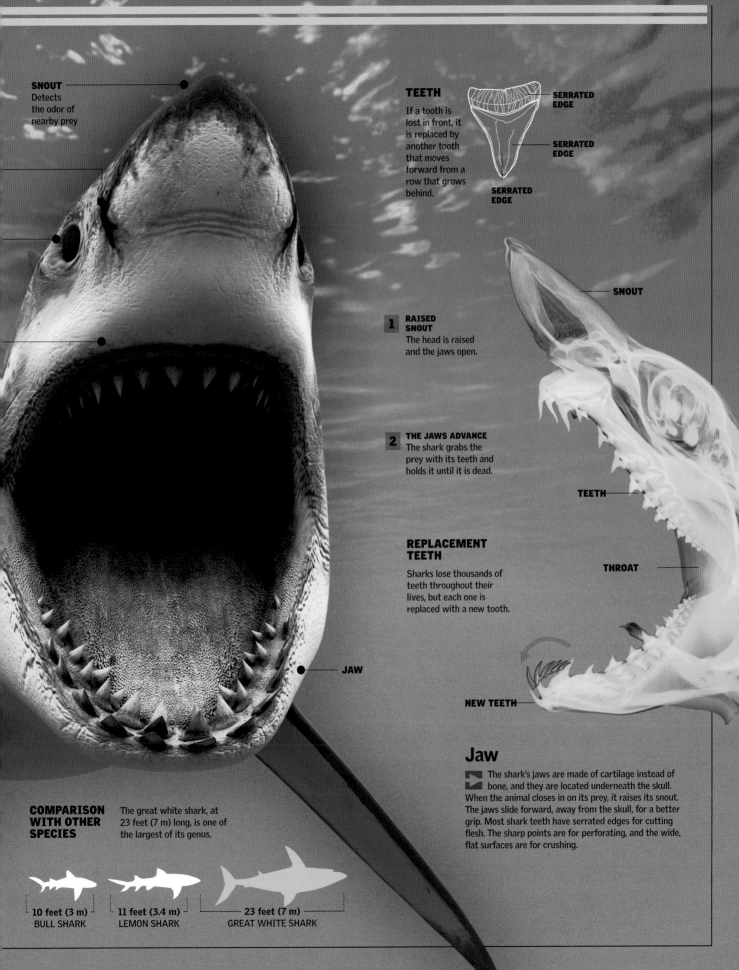

SNOUT
Detects
the odor of
nearby prey

TEETH
If a tooth is
lost in front, it
is replaced by
another tooth
that moves
forward from a
row that grows
behind.

SERRATED
EDGE

SERRATED
EDGE

SERRATED
EDGE

SNOUT

**1 RAISED
SNOUT**
The head is raised
and the jaws open.

2 THE JAWS ADVANCE
The shark grabs the
prey with its teeth and
holds it until it is dead.

TEETH

**REPLACEMENT
TEETH**

Sharks lose thousands of
teeth throughout their
lives, but each one is
replaced with a new tooth.

THROAT

JAW

NEW TEETH

Jaw

The shark's jaws are made of cartilage instead of
bone, and they are located underneath the skull.
When the animal closes in on its prey, it raises its snout.
The jaws slide forward, away from the skull, for a better
grip. Most shark teeth have serrated edges for cutting
flesh. The sharp points are for perforating, and the wide,
flat surfaces are for crushing.

**COMPARISON
WITH OTHER
SPECIES**

The great white shark, at
23 feet (7 m) long, is one of
the largest of its genus.

10 feet (3 m)
BULL SHARK

11 feet (3.4 m)
LEMON SHARK

23 feet (7 m)
GREAT WHITE SHARK

The Journey Home

After living in the ocean for five or six years, the Pacific red salmon (*Oncorhynchus nerka*) returns to the river where it was born to reproduce. The journey lasts from two to three months, and it demands a great deal of energy. The salmon must swim against the current, climb waterfalls, and evade predators, including bears and eagles. Once the salmon reach the river, the female lays her eggs, and the male fertilizes them. Typically, the same locations in specific rivers are sought year after year. This species of salmon dies after completing the reproductive cycle, unlike the Atlantic salmon, which repeats the cycle three or four times. Once the eggs hatch, the cycle begins anew.

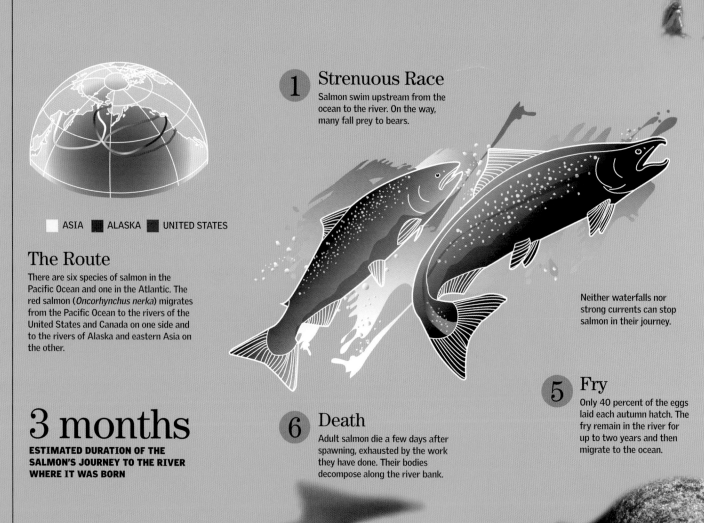

ASIA ALASKA UNITED STATES

The Route

There are six species of salmon in the Pacific Ocean and one in the Atlantic. The red salmon (*Oncorhynchus nerka*) migrates from the Pacific Ocean to the rivers of the United States and Canada on one side and to the rivers of Alaska and eastern Asia on the other.

1 Strenuous Race

Salmon swim upstream from the ocean to the river. On the way, many fall prey to bears.

Neither waterfalls nor strong currents can stop salmon in their journey.

3 months

ESTIMATED DURATION OF THE SALMON'S JOURNEY TO THE RIVER WHERE IT WAS BORN

6 Death

Adult salmon die a few days after spawning, exhausted by the work they have done. Their bodies decompose along the river bank.

5 Fry

Only 40 percent of the eggs laid each autumn hatch. The fry remain in the river for up to two years and then migrate to the ocean.

2 Red River

The salmon returns to its birthplace to spawn. Males have intense coloration with a green head.

Survival

Of the more than 7,500 eggs that two females can lay, only two hatched fish will remain at the end of the life cycle of two years. Many eggs die before hatching, and after hatching, salmon fry are easy prey for other fish.

EGGS	7,500
FRY	4,500
FRY	650
FRY	200
SALMON	50
ADULT SALMON	4
EGG SPAWNING	2

6–Year Cycle
LENGTH OF TIME FROM SPAWNING TO ADULTHOOD

Seen from above, salmon appear as a large red spot.

3 The Couple

While females are busy preparing nests in the sand to deposit their eggs, males compete for mates.

BACK
A hump develops in the dorsal section of the body.

5,000
QUANTITY OF EGGS A FEMALE CAN LAY

MOUTH
During mating season, the lower jaw of the male curves upward.

COLOR
The blue-backed salmon turns a fiery red.

4 Spawning

The female deposits between 2,500 and 5,000 eggs in a series of nests. The male fertilizes them as they fall between the rocks.

Habitat, Tastes, and Preferences

The oceans cover 70 percent of the earth's surface. That is where life began on this planet and where the most primitive species live side by side with the most highly evolved ones. This abundance of species is due in part to the wide variety of environments found in the ocean. As one descends in depth, the water's temperature decreases, as does the amount of light. These factors determine different ecosystems, feeding patterns, and adaptation strategies among a wide variety of fish species.

Reserve of Life

Corals need warm water and a lot of light. They are colonies of polyps that secrete a calcareous substance that form great reefs over the years. The resulting microhabitat harbors a great variety of species.

REEFS
form in only very shallow tropical waters.

0–650 feet
(0–200 m)
EPIPELAGIC ZONE
Algae and the animals that eat them inhabit this zone, where photosynthesis is possible because of the presence of sunlight.

490 feet
(150 m)
At this depth, there is no plankton. Many species that live below this depth swim above it at night to feed.

650 FEET (200 M)

30 FEET (9 M)
Divers without special equipment

50 FEET (15 M)
Pearl divers

160 FEET (50 M)
Scuba divers

PLANKTON
Herbivorous fish can be found only in shallow waters because of the presence of plankton there.

SWORDFISH

FLYING FISH

TRUMPET FISH

CLOWNFISH

HAMMERHEAD SHARK

MANTA RAY

SNAPPER

TUNA

COD

SERGEANT FISH

BLUE ANGELFISH

STRIPED PERCH

BARRACUDA

PACIFIC SARDINES

MORAY EEL

OCEAN SUNFISH

PUFFER FISH

TIGER SHARK

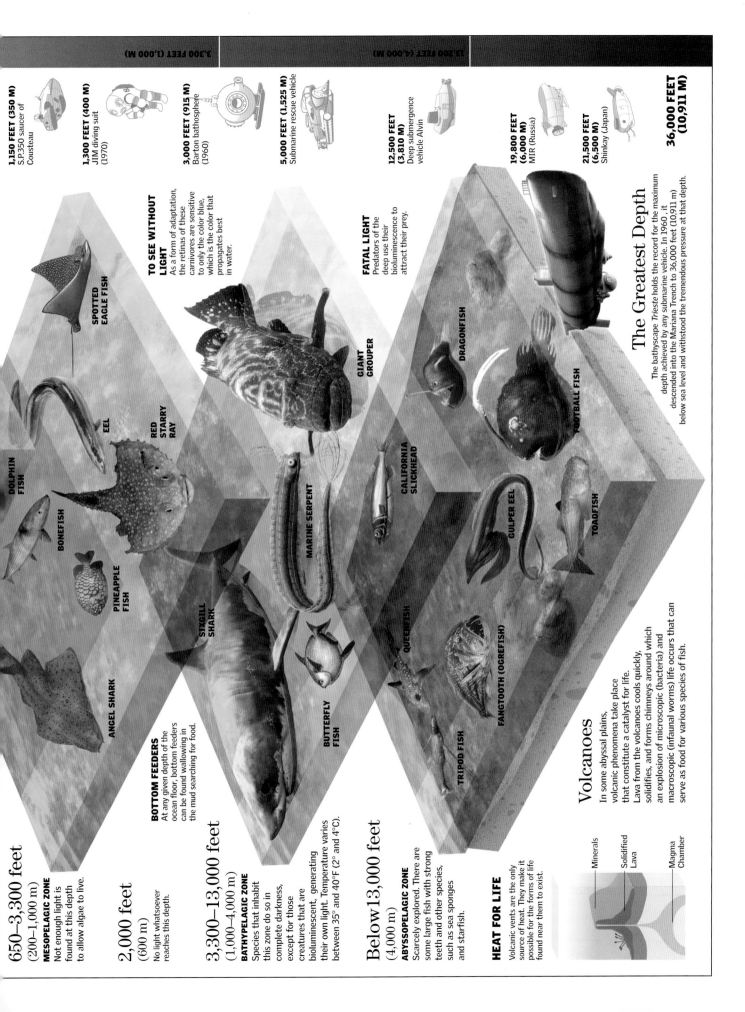

3,300 FEET (1,000 M)

13,200 FEET (4,000 M)

1,150 FEET (350 M)
S.P.350 saucer of Cousteau

1,300 FEET (400 M)
JIM diving suit (1970)

3,000 FEET (915 M)
Barton bathosphere (1960)

5,000 FEET (1,525 M)
Submarine rescue vehicle

12,500 FEET (3,810 M)
Deep submergence vehicle Alvin

19,800 FEET (6,000 M)
MIR (Russia)

21,500 FEET (6,500 M)
Shinkay (Japan)

36,000 FEET (10,911 M)

650–3,300 feet
(200–1,000 m)

MESOPELAGIC ZONE
Not enough light is found at this depth to allow algae to live.

2,000 feet
(600 m)
No light whatsoever reaches this depth.

3,300–13,000 feet
(1,000–4,000 m)

BATHYPELAGIC ZONE
Species that inhabit this zone do so in complete darkness, except for those creatures that are bioluminescent, generating their own light. Temperature varies between 35° and 40°F (2° and 4°C).

Below 13,000 feet
(4,000 m)

ABYSSOPELAGIC ZONE
Scarcely explored. There are some large fish with strong teeth and other species, such as sea sponges and starfish.

HEAT FOR LIFE
Volcanic vents are the only source of heat. They make it possible for the forms of life found near them to exist.

BOTTOM FEEDERS
At any given depth of the ocean floor, bottom feeders can be found wallowing in the mud searching for food.

TO SEE WITHOUT LIGHT
As a form of adaptation, the retinas of these carnivores are sensitive to only the color blue, which is the color that propagates best in water.

FATAL LIGHT
Predators of the deep use their bioluminescence to attract their prey.

The Greatest Depth
The bathyscape *Trieste* holds the record for the maximum depth achieved by any submarine vehicle. In 1960, it descended into the Mariana Trench to 36,000 feet (10,911 m) below sea level and withstood the tremendous pressure at that depth.

Volcanoes
In some abyssal plains, volcanic phenomena take place that constitute a catalyst for life. Lava from the volcanoes cools quickly, solidifies, and forms chimneys around which an explosion of microscopic (bacteria) and macroscopic (infaunal worms) life occurs that can serve as food for various species of fish.

Minerals

Solidified Lava

Magma Chamber

SPOTTED EAGLE FISH

EEL

DOLPHIN FISH

BONEFISH

PINEAPPLE FISH

ANGEL SHARK

RED STARRY RAY

GIANT GROUPER

MARINE SERPENT

SIXGILL SHARK

BUTTERFLY FISH

CALIFORNIA SLICKHEAD

DRAGONFISH

FOOTBALL FISH

GULPER EEL

TOADFISH

QUEENFISH

FANGTOOTH (OGREFISH)

TRIPOD FISH

Kings of Darkness

In depths below 8,200 feet (2,500 m), where barely any light penetrates, live rare species known as abyssal fish. In this environment, life is possible near hydrothermal vents in the seafloor that warm the nearby waters. In spite of this natural warmth, in many areas the temperature never rises above 36°F (2°C). At this depth, fish have peculiar shapes, with large heads and strong teeth for eating other fish, because no vegetation can grow there. To attract their prey, many have "lure" organs made of photophores that shine in the darkness. They also are usually black or dark brown for purposes of camouflage.

FANFIN SEADEVIL
Caulophryne jordani
This dark-brown fish uses the photophore organ on its head to penetrate the darkness.

SLOANE'S VIPERFISH
Chauliodus sloani
Between 12 and 20 inches (30 and 50 cm) in stature, it is dark blue or silvery in color and lives in warm tropical waters.

SHARP, POINTED TEETH
It gulps down its prey after grabbing them with its enormous teeth and its strong suction force.

LANTERN
Like most abyssal fish, it has a lure organ.

EYES FOR SEEING IN DIM LIGHT

RAY OF LIGHT

TAPETUM
reflects light like a mirror. Each ray hits the retina twice, doubling its sensitivity.

RETINA
Blind to red light. It registers only blue light waves, which travel better in the water.

FANGTOOTH
Anoplogaster cornuta
This fearsome hunter kills its prey by seizing it with its jaw and strong teeth.

FILAMENTS
cover its entire body for protection.

DRAGONFISH
Bathophilus sp.
Found in most tropical regions of the world, it has photophores along both sides of its body.

CHIN APPENDAGE
Shines in the darkness

FUMAROLE
Openings in the earth's surface that discharge geothermal water and minerals. As the water cools, these minerals solidify.

36°F (2°C)

TEMPERATURE OF WATER HEATED BY FUMAROLES

TUBE WORM TENTACLES
Tube worms have neither mouths nor digestive tracts. They feed on organic molecules formed from elements in the water by chemosynthetic bacteria that live inside the worms.

LANTERN
produces bluish light, which reaches farthest underwater.

GLOWING LURE
Produces light to attract prey

DIMENSIONS

Weight
10½ ounces
(300 g)

4 inches (10 cm)

HUMPBACK ANGLERFISH
Melanocetus johnsonii
6 inches (15 cm) long. Its small fins are insufficient to enable fast maneuvering.

SKIN
Dark colors are likely to make it invisible to attackers.

HYDROSTATIC PRESSURE
The weight of the column of water. The pressure of the water increases with depth. In the Mariana Trench (the deepest undersea trench on the planet), every square inch bears the weight of 7¾ tons (1¼ tons per square centimeter) of water.

1.3 cubic yards (1 cu m) of water = 1.1 tons (1,000 kg)

8,200 feet (2,500 m)

DEPTH OF WATER

KILLER JAWS
In the ocean's depths, only the best hunter survives.

CHIN APPENDAGE
Produces light to attract prey

ILLUMINATED NETDEVIL
Linophryne arborifera
has a glowing lure on the end of its nose and a branching beard that also glows to attract prey. The male is smaller than the female and lives off of her like a parasite.

GLOWING LURE
gives off light to attract prey.

TAILS AND FINS
contain luminous cells.

ATLANTIC FOOTBALL FISH
Himantolophus groenlandicus
The females can reach up to 24 inches (60 cm) long, whereas the males barely reach 1½ inches (4 cm) long and live as parasites on their mates.

Sea Snakes

Eels (Anguilliformes), an order of ray-finned fish (Actinopterygii), are distinctive for their elongated, snakelike shape. In the past, they were an important food source. There are about 600 species of true eels, including morays, congers, and snake eels. Eels come in a wide variety of colors and patterns, ranging from solid gray to mottled yellow. Their bodies lack scales and are covered with a protective mucous membrane. One of the most striking eels is the green moray, which lives in the Caribbean Sea and hides in coral reefs awaiting its prey. Although it is not poisonous, it is feared by divers because its bite can inflict grave wounds.

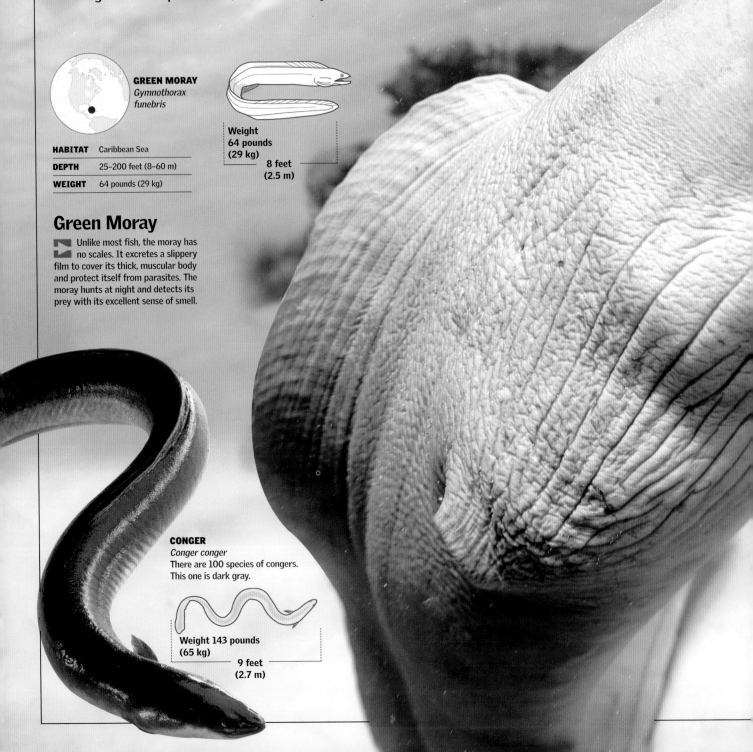

GREEN MORAY
Gymnothorax funebris

Weight
64 pounds
(29 kg)

8 feet
(2.5 m)

HABITAT	Caribbean Sea
DEPTH	25–200 feet (8–60 m)
WEIGHT	64 pounds (29 kg)

Green Moray

Unlike most fish, the moray has no scales. It excretes a slippery film to cover its thick, muscular body and protect itself from parasites. The moray hunts at night and detects its prey with its excellent sense of smell.

CONGER
Conger conger
There are 100 species of congers. This one is dark gray.

Weight 143 pounds
(65 kg)

9 feet
(2.7 m)

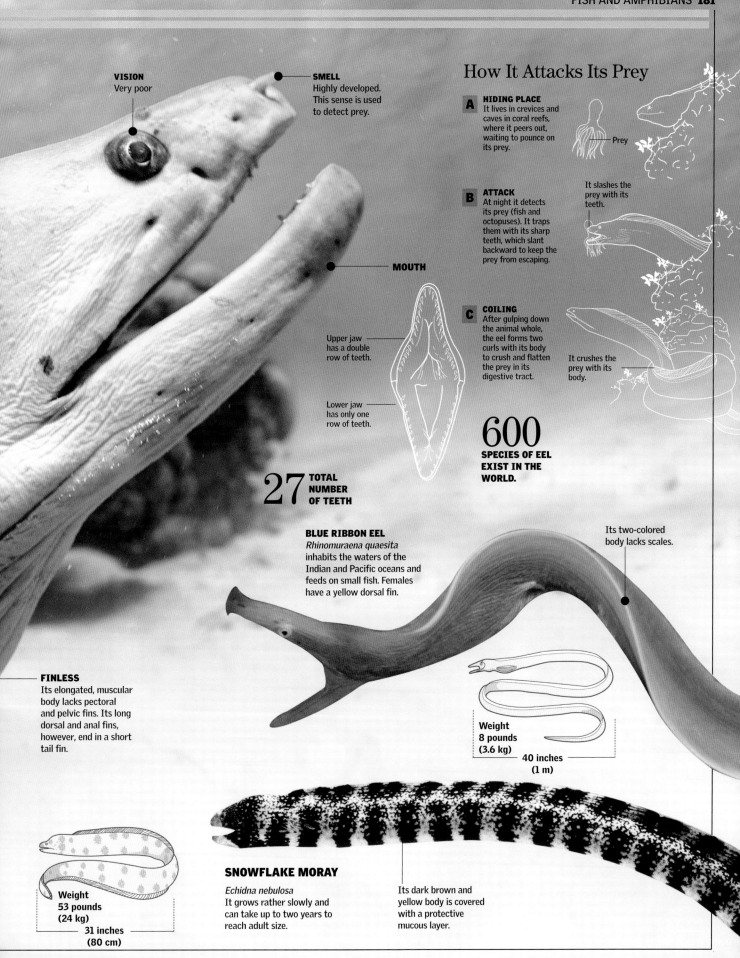

VISION
Very poor

SMELL
Highly developed.
This sense is used
to detect prey.

How It Attacks Its Prey

A HIDING PLACE
It lives in crevices and
caves in coral reefs,
where it peers out,
waiting to pounce on
its prey.

Prey

B ATTACK
At night it detects
its prey (fish and
octopuses). It traps
them with its sharp
teeth, which slant
backward to keep the
prey from escaping.

It slashes the
prey with its
teeth.

C COILING
After gulping down
the animal whole,
the eel forms two
curls with its body
to crush and flatten
the prey in its
digestive tract.

It crushes the
prey with its
body.

MOUTH

Upper jaw
has a double
row of teeth.

Lower jaw
has only one
row of teeth.

27 TOTAL
NUMBER
OF TEETH

600
SPECIES OF EEL
EXIST IN THE
WORLD.

BLUE RIBBON EEL
Rhinomuraena quaesita
inhabits the waters of the
Indian and Pacific oceans and
feeds on small fish. Females
have a yellow dorsal fin.

Its two-colored
body lacks scales.

FINLESS
Its elongated, muscular
body lacks pectoral
and pelvic fins. Its long
dorsal and anal fins,
however, end in a short
tail fin.

**Weight
8 pounds
(3.6 kg)**
**40 inches
(1 m)**

**Weight
53 pounds
(24 kg)**
**31 inches
(80 cm)**

SNOWFLAKE MORAY

Echidna nebulosa
It grows rather slowly and
can take up to two years to
reach adult size.

Its dark brown and
yellow body is covered
with a protective
mucous layer.

Amphibians

F ew groups of amphibians have generated as much scientific interest as frogs of the genus *Dendrobates*, which produce toxic secretions through their skin. All frogs of this genus have spectacular coloring to warn their predators of the danger. One of the most important traits of amphibians (newts, salamanders, frogs,

POISONOUS FROG
Frogs of the genus
Dendrobates secrete a
special type of poison that
attacks the nervous system.

toads, and caecilians) has been their conquest of land. This completely transformed the extremities of these animals, allowing them to move on land instead of swimming. They also had to adapt to take in oxygen through their skin and lungs. Here you will also discover how frogs and toads reproduce and how newts feed, among other curious facts.

Between Land and Water

As indicated by their name (amphi, "both," and bios, "life"), these animals lead a double life. When young, they live in the water, and when they become adults they live outside it. In any case, many must remain near water or in very humid places to keep from drying out. This is because amphibians also breathe through their skin, and only moist skin can absorb oxygen. Some typical characteristics of adult frogs and toads include a tailless body, long hind limbs, and large eyes that often bulge.

Amphibian Anatomy

Amphibian anatomy has several peculiarities. Larvae, such as tadpoles, have a respiratory system with gills. Most species develop lungs when they reach adulthood. They also have a trachea, pharynx, and saclike lungs, even though skin breathing is at times more important than lung breathing. The heart has two auricles and one ventricle, and the digestive and excretory systems are similar to those of mammals.

The Skin

Amphibians breathe through their skin, which is clean and smooth, without hair or scales. They must always keep it moist, because it has a strong tendency to dry out. Even though they have mucous glands that help maintain moisture, amphibians must live in damp places. The skin of most amphibians protects them from possible predators and has poisonous glands that secrete unpleasant and even toxic substances.

HIND LIMBS
The muscular leg and foot have five long fingers joined by a webbed membrane that aids swimming.

VOCAL SACS

Both toads and frogs sing. Even though the sound is produced by their vocal cords, in males the sound is amplified by means of inflatable sacs on each side of the larynx.

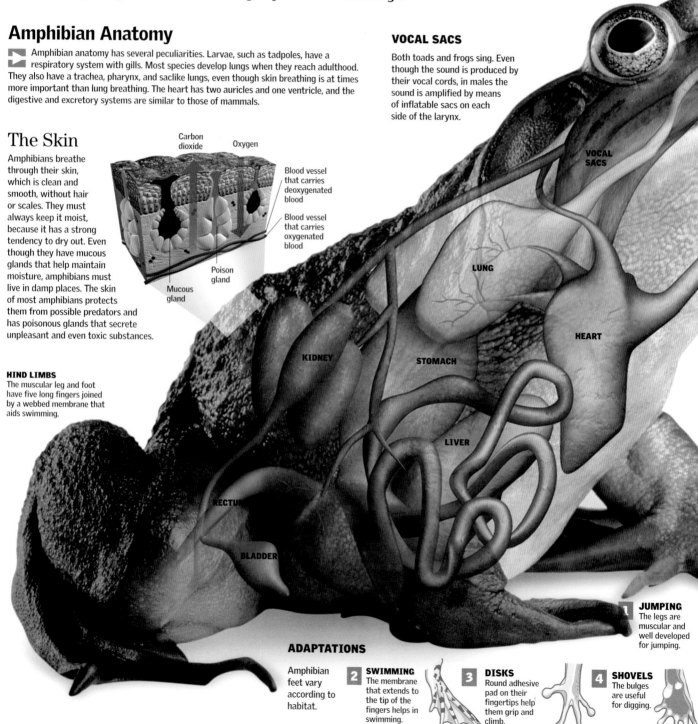

Carbon dioxide

Oxygen

Blood vessel that carries deoxygenated blood

Blood vessel that carries oxygenated blood

Poison gland

Mucous gland

VOCAL SACS

LUNG

HEART

KIDNEY

STOMACH

LIVER

RECTUM

BLADDER

ADAPTATIONS

Amphibian feet vary according to habitat.

2 SWIMMING
The membrane that extends to the tip of the fingers helps in swimming.

3 DISKS
Round adhesive pad on their fingertips help them grip and climb.

1 JUMPING
The legs are muscular and well developed for jumping.

4 SHOVELS
The bulges are useful for digging.

Differences Between Frogs and Toads

It is very common to use "frog" and "toad" as synonyms or to think that the frog is a female toad. However, frogs and toads are quite different. Toads have wrinkled skin and short legs, and they are land animals. Frogs are smaller, have webbed feet, and live in the water and in trees.

SKIN
Soft and smooth, with strong, bright colors

EYES
Frogs have horizontal pupils.

EYES
The pupil is usually horizontal, although some toads have vertical pupils.

SKIN
The skin of a toad is wrinkled, hard, rough, and dry. It is also used as leather.

COMMON TOAD
Bufo bufo

REED FROG
Hyperolius tuberilinguis

POSTURE
Toads are terrestrial species, slow-moving and wider than frogs. Frogs live mainly in water, which is why they have webbed toes adapted for swimming.

LEGS
are long and are adapted for jumping. Frogs have webbed toes to help with swimming.

LEGS
are shorter and wider than those of frogs and are adapted for walking.

CATCHING
Toads gulp down their prey, swallowing it whole.

SWALLOWING
Eye retraction, where the toad closes and turns its eyes inward, increases the pressure in the mouth, pushing food down the esophagus.

Nutrition

is based on plants during the larval stage, whereas in the adult stage the main food sources are arthropods (such as insects of the order Coleoptera and arachnids) and other invertebrates, such as butterfly caterpillars and earthworms.

Types of Amphibians

Amphibians are divided into three groups that are differentiated on the basis of tail and legs. Newts and salamanders have tails. They belong to the order Urodela. Frogs and toads, which have no tail except as tadpoles, belong to the Anura group. Caecilians, which have no tail or legs, are similar to worms and belong to the Apoda group.

1 ANURA
Tailless

EUROPEAN TREE FROG
is docile and lives near buildings.

2 APODA
Without legs

RINGED CAECILIAN
looks like a large, thick worm.

3 URODELA
With a tail

TIGER SALAMANDER
One of the most colorful in America

Legs

Frogs and toads have four fingers on each front leg and five on each hind leg. Water frogs have webbed feet; tree frogs have adhesive disks on the tips of their fingers to hold on to vertical surfaces; and burrowing frogs have callous protuberances called tubercules on their hind legs, which they use for digging.

Jumping Athletes

Amphibians of the order Anura are known for their ability to jump high and far. This group includes frogs and toads, and their anatomy helps them to jump. Frogs use their jumping ability to escape from their many predators; they can jump a distance equivalent to 10 to 44 times their body length. When they feel threatened, they can choose to jump into the nearest body of water, where they hide, or they can jump erratically on land to confuse their attacker.

② Feeding

Amphibians from the order Anura have a varied diet. They feed on insects and small invertebrates, such as earthworms, snails, crustaceans, and spiders. Tadpoles are herbivores.

The Frog

Its large eyes help it to locate prey easily. The eyes have lids that protect them from particles in the air or help them see underwater. The frog's smooth skin has glands that moisten it or that secrete toxic or irritating substances. The frog breathes through its lungs and skin. It has a large tympanum, or eardrum, visible on each side of the head, and a wide mouth that may or may not have teeth.

① Jumping

Before the jump begins, the frog tenses the muscles of its hind legs and presses its feet against the ground. As the frog jumps, the legs extend to propel the body forward.

VISIBLE HUMP

STRETCHED-OUT BODY

LEG MUSCLES
tense to carry out the jump.

EDIBLE FROG
Rana esculenta
is found in Europe and also in the United States, Canada, and Asia.

HIND FEET
have five webbed toes.

The Toad

Having characteristics similar to those of frogs, toads can be distinguished by only a few features. Generally, toads are larger, less stylized, and better adapted to living on land. Toads' skin is thicker than that of frogs to prevent it drying, and toads are normally covered with warts.

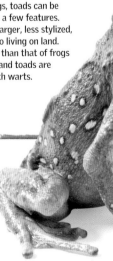

ASIAN TREE FROG
Pedostibes tuberculosus

HOW IT FEEDS

EYES
During the jump the eyes remain shut.

1 ADHESION
The insect adheres to the tip of the tongue, which is sticky.

2 NO ESCAPE
The tongue folds back into the mouth, carrying the insect with it.

Insects found on plants are the favorite meal of frogs.

FOREFEET
have four fingers and are not as strong as the hind feet.

AT THE TOP

The white-lipped tree frog (*Litoria infrafrenata*) has a maximum length of 4 inches (10 cm) and is adapted for mountain climbing, jumping, and moving on flat areas. The pads on the tip of each finger and toe allow it to adhere to many surfaces.

The toe has a sticky mucous coating.

WHITE-LIPPED TREE FROG
Litoria infrafrenata

SPINAL COLUMN
Its small number of vertebrae give it elasticity in the act of jumping.

9 VERTEBRAE
Aside from these, it has a urostyle—a cylindrical bone that results from the fusion of vertebrae.

LEG
Adapted for jumping and swimming

3

Fall

At this moment, when the frog extends its hind legs, it not only reduces air resistance but also helps with the entry into the water.

FALL
Each hind leg extends like an arrow.

$17\frac{1}{2}$ feet
(5.35 m)
THE DISTANCE JUMPED BY AN AFRICAN FROG

DIVE
The body curves upward when it enters the water.

THE JUMP

The toad's jump covers less distance because of its greater weight and because its legs are not as flexible as those of the frog.

The hind legs boost the jump.

It closes its eyes for protection.

It lands with its forelimbs.

It rises in the air a few inches.

AT REST **DRIVE** **FULL JUMP** **LANDING**

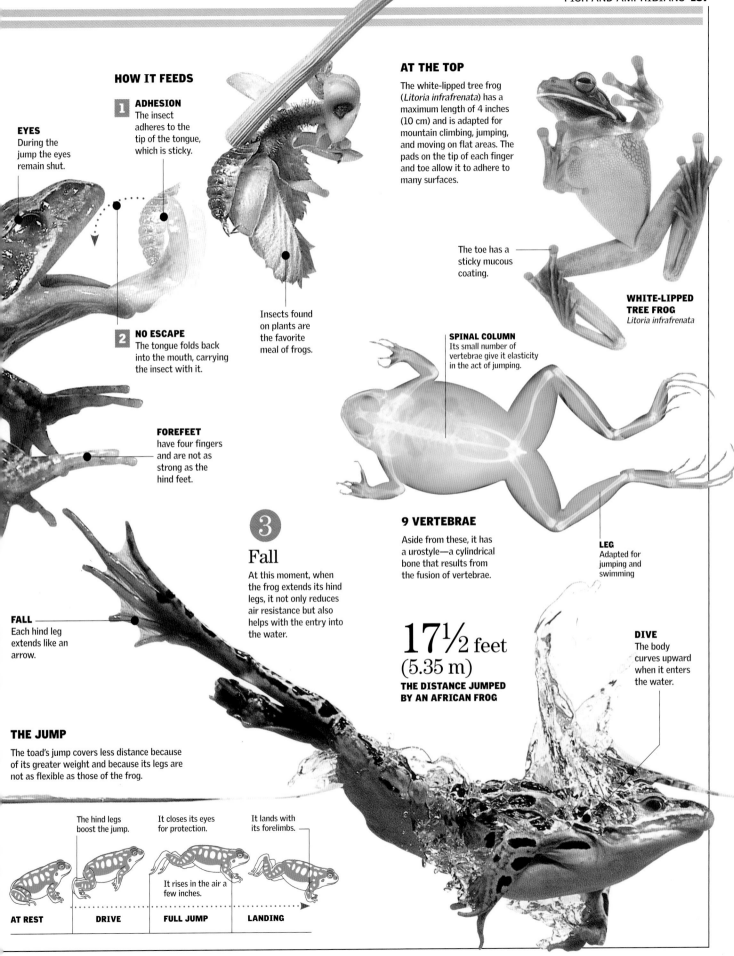

Deep Embrace

Reproduction by amphibians is usually carried out in the water, where the female deposits the eggs, despite the fact that some species are able to deposit eggs on land. The most favorable time for this activity is during the spring, when the male sings to make his presence known. During mating, also called amplexus, the male positions himself on top and fertilizes the eggs as they come out. Then gelatinous layers absorb water and increase their volume, binding the eggs together in large masses.

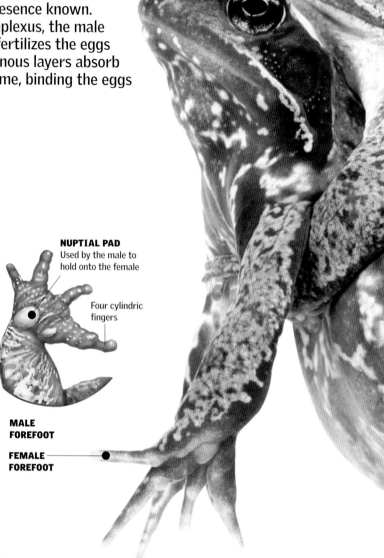

A ROMANTIC SONG
The call that a male makes to mate with a female

Amplexus

Fertilization for the majority of amphibians is external. In this hazardous process, the male, embracing the female in amplexus, discharges spermatozoa while the ovocytes are released. Both are released in great numbers in order to ensure the success of the process. This mating embrace can last from 23 to 45 minutes.

NUPTIAL PAD
Used by the male to hold onto the female

Four cylindric fingers

MALE FOREFOOT

FEMALE FOREFOOT

─2³/₄ inches─
(7 cm)

Females are larger than males.

Weight
1³/₄–5 ounces
(50–100 g)

IBERIAN WATER FROG
Rana perezi

DIET	Carnivorous
REPRODUCTION	Oviparous
SEASON	Spring

EGGS INSIDE THE FEMALE

SOME ANURANS CAN LAY UP TO

20,000 eggs.

LIFE CYCLE

The three stages of the life cycle are egg, larva, and adult. The embryos begin to develop within the eggs; then, after six or nine days, the eggs hatch, and tiny tadpoles with spherical heads, large tails, and gills emerge. Once the gills pass their function over to the lungs and the tail of the amphibian has shrunk and disappeared, the young frog enters the adult stage.

Sperm   Oocyte

SPERM
Male gamete

Plasma germ cell

OOCYTE
Female gamete

Sexually mature male

Gonad

Zygote   Morula   Blastula

Blastocele

16 weeks
THE DURATION OF THE CYCLE

Blastopore

Ectoderm   Mesoderm

Ectoderm

Immature larva   Birth

Responsible Parents

Some males of frog and toad species play an important role in the protection of the eggs laid by the female. They pick up the eggs and help the mothers, and some even carry the eggs with them until the birth takes place.

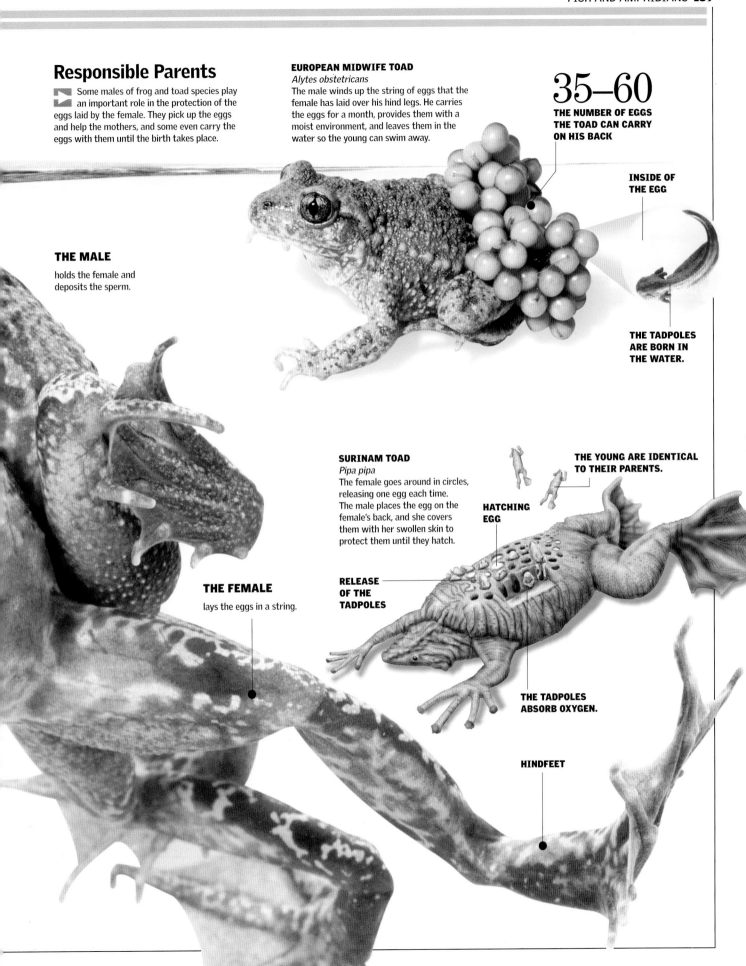

EUROPEAN MIDWIFE TOAD
Alytes obstetricans
The male winds up the string of eggs that the female has laid over his hind legs. He carries the eggs for a month, provides them with a moist environment, and leaves them in the water so the young can swim away.

35–60
THE NUMBER OF EGGS THE TOAD CAN CARRY ON HIS BACK

INSIDE OF THE EGG

THE TADPOLES ARE BORN IN THE WATER.

THE MALE
holds the female and deposits the sperm.

SURINAM TOAD
Pipa pipa
The female goes around in circles, releasing one egg each time. The male places the egg on the female's back, and she covers them with her swollen skin to protect them until they hatch.

THE YOUNG ARE IDENTICAL TO THEIR PARENTS.

HATCHING EGG

THE FEMALE
lays the eggs in a string.

RELEASE OF THE TADPOLES

THE TADPOLES ABSORB OXYGEN.

HINDFEET

Metamorphosis

Metamorphosis is the process of transformation experienced by anurans (it can also be observed in amphibians from the order Urodela and caecilians), starting with the egg and ending at the adult stage. When they leave the egg, amphibians have a larval form. They then undergo very important changes in their anatomy, diet, and lifestyle, slowly mutating from their first stage, which is completely aquatic, until they transform into animals adapted to life on land.

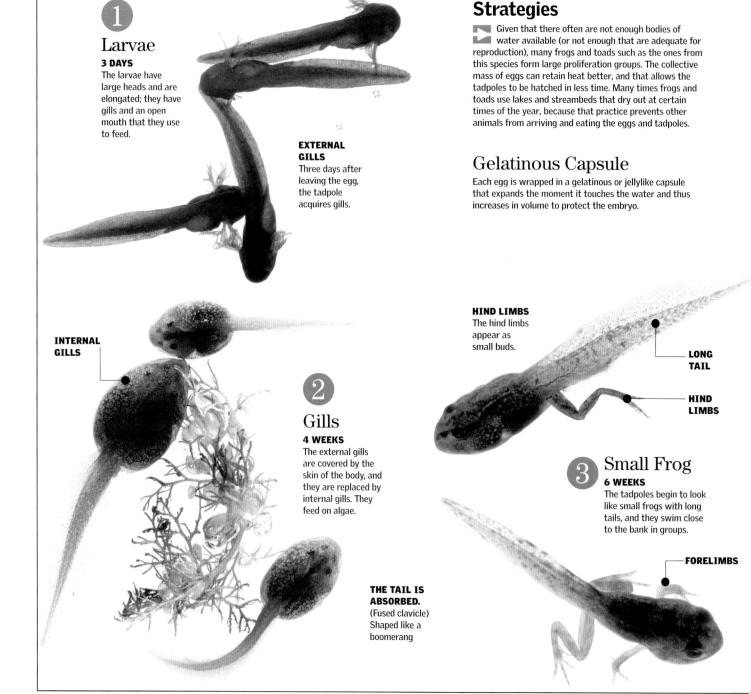

1 Larvae

3 DAYS
The larvae have large heads and are elongated; they have gills and an open mouth that they use to feed.

EXTERNAL GILLS
Three days after leaving the egg, the tadpole acquires gills.

INTERNAL GILLS

2 Gills

4 WEEKS
The external gills are covered by the skin of the body, and they are replaced by internal gills. They feed on algae.

THE TAIL IS ABSORBED.
(Fused clavicle) Shaped like a boomerang

Strategies

Given that there often are not enough bodies of water available (or not enough that are adequate for reproduction), many frogs and toads such as the ones from this species form large proliferation groups. The collective mass of eggs can retain heat better, and that allows the tadpoles to be hatched in less time. Many times frogs and toads use lakes and streambeds that dry out at certain times of the year, because that practice prevents other animals from arriving and eating the eggs and tadpoles.

Gelatinous Capsule

Each egg is wrapped in a gelatinous or jellylike capsule that expands the moment it touches the water and thus increases in volume to protect the embryo.

HIND LIMBS
The hind limbs appear as small buds.

LONG TAIL

HIND LIMBS

3 Small Frog

6 WEEKS
The tadpoles begin to look like small frogs with long tails, and they swim close to the bank in groups.

FORELIMBS

Cycle
METAMORPHOSIS

The development of the common European frog from egg to adult takes approximately 16 weeks.

Mother Frog and Her Eggs

Despite the fact that the survival instinct of anurans is not fully developed, frogs and toads somehow take care of their future young. Laying eggs in great quantities ensures that many tadpoles will be able to escape predators who feed on the eggs. The gelatinous layer also protects the eggs from other predators. Some frogs even care for their tadpoles by nestling them on their backs. An example of such a frog is the Surinam toad.

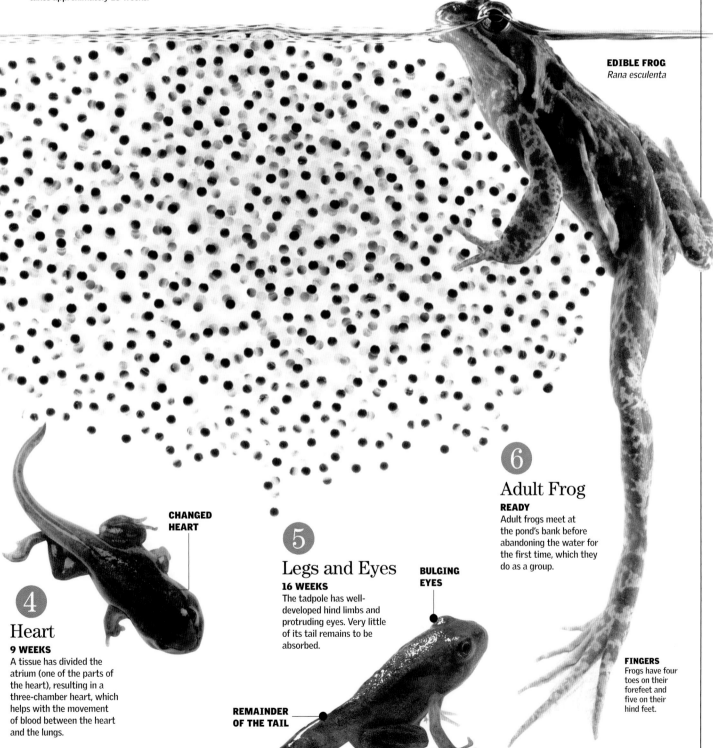

EDIBLE FROG
Rana esculenta

CHANGED HEART

6
Adult Frog
READY
Adult frogs meet at the pond's bank before abandoning the water for the first time, which they do as a group.

5
Legs and Eyes
16 WEEKS
The tadpole has well-developed hind limbs and protruding eyes. Very little of its tail remains to be absorbed.

BULGING EYES

4
Heart
9 WEEKS
A tissue has divided the atrium (one of the parts of the heart), resulting in a three-chamber heart, which helps with the movement of blood between the heart and the lungs.

REMAINDER OF THE TAIL

FINGERS
Frogs have four toes on their forefeet and five on their hind feet.

A Very Peculiar Tail

The salamander is an animal of the order Urodela that needs damp places to survive. It lives in a very limited range of areas, and it is highly sensitive to modifications in its natural habitat. Unlike frogs and toads, the salamander keeps its tail when it reaches adulthood. The tail makes up nearly half the length of its body. Salamanders, especially adults, are completely nocturnal. Their movements are slow when they walk or crawl along the ground. During the day, they stay hidden under rocks, in underground burrows, and on tree trunks.

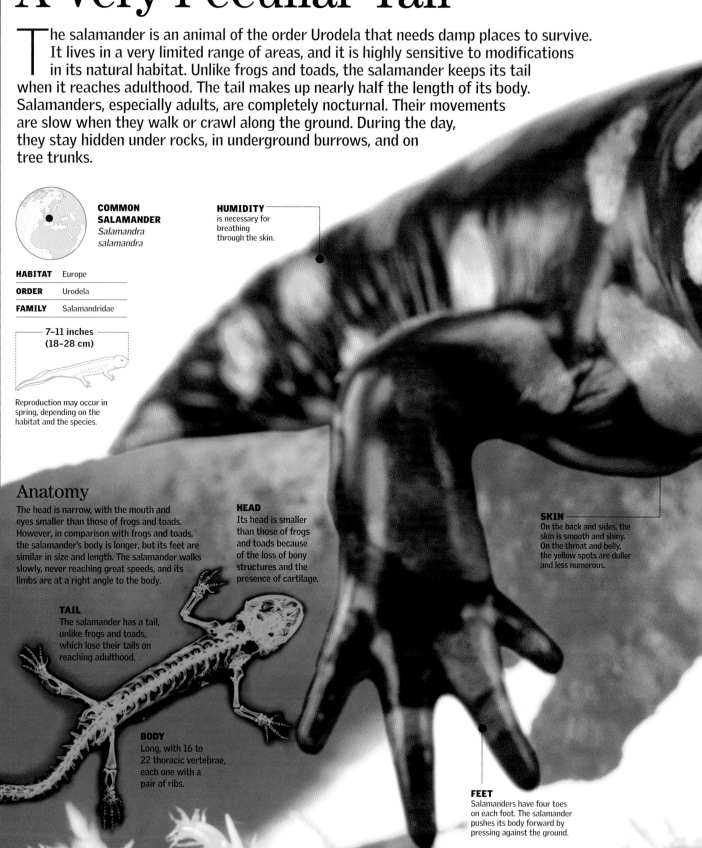

COMMON SALAMANDER
Salamandra salamandra

HABITAT	Europe
ORDER	Urodela
FAMILY	Salamandridae

7–11 inches (18–28 cm)

Reproduction may occur in spring, depending on the habitat and the species.

HUMIDITY
is necessary for breathing through the skin.

Anatomy

The head is narrow, with the mouth and eyes smaller than those of frogs and toads. However, in comparison with frogs and toads, the salamander's body is longer, but its feet are similar in size and length. The salamander walks slowly, never reaching great speeds, and its limbs are at a right angle to the body.

HEAD
Its head is smaller than those of frogs and toads because of the loss of bony structures and the presence of cartilage.

SKIN
On the back and sides, the skin is smooth and shiny. On the throat and belly, the yellow spots are duller and less numerous.

TAIL
The salamander has a tail, unlike frogs and toads, which lose their tails on reaching adulthood.

BODY
Long, with 16 to 22 thoracic vertebrae, each one with a pair of ribs.

FEET
Salamanders have four toes on each foot. The salamander pushes its body forward by pressing against the ground.

EYE
Large and bulging, with a dark brown iris

TONGUE PAD

THE TONGUE MUSCLES RETRACT.

OUTER SECTION OF THE TONGUE

RETRACTOR MUSCLES

Feeding Habits

With its long tongue, the salamander can trap its prey in a flash and quickly gulp it down. These carnivorous animals use mainly sight and smell to hunt. Because they are not very active, salamanders need relatively small amounts of food. If they obtain more food than necessary, they store it as fat for lean times.

Life Cycle

There are three stages to the life cycle: egg, larva, and adult. The eggs vary in size depending on the species. Larvae have feathery external gills. Metamorphosis lasts until adulthood, when the salamander loses its gills and switches to breathing with lungs.

1 EGG
Hatches into a larva

Defense

The Italian spectacled salamander has two ways of avoiding its enemies. It plays dead, or it curls its tail forward. Other species defend themselves by using a toxic substance produced by glands or by breaking off the tail, which continues to move on its own and confuses the predator.

ITALIAN SALAMANDER

2 BIRTH
The larva is born with feathery external gills.

55 years
LIFE SPAN OF SOME SPECIES

3 ADULT
Metamorphosis is completed; the salamander reaches sexual maturity.

CHANGE
The body grows longer; the salamander begins to breathe through the skin and lungs.

LARVA
Metamorphosis begins; the salamander loses its gills and switches to breathing air.

LARGE ALPINE SALAMANDER
Salamandra lanzai
is known for having the longest gestational period of all animals, even longer than that of elephants.

38 months
GESTATIONAL PERIOD

Newts

Along with salamanders, newts are the most primitive of terrestrial vertebrates. Of the three main surviving groups of primitive amphibians, newts most closely resemble the animals from which all amphibians are descended. Some of their habits are also more complex and varied. Most of the time they live on land, but during the mating season they return to the water. Unlike frogs and toads, newts and salamanders keep their tails as adults. They are found in temperate regions of the Northern Hemisphere.

FRONT FEET
Newts have four toes on each of their front feet.

Courtship and Reproduction

Courtship and mating involve a showy exhibition by both male and female. The male must find a female of the same species and bring her a packet of sperm, which he deposits on the ground or in a pool. Fertilization is internal, and the female gathers the packet into her cloaca.

1 DANCE
Males are attracted by the female's belly, swollen with eggs. The males draw her attention with their showy pigmentation and the flexible crest along their back and tail.

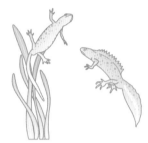

2 EXHIBITION
The male swims in front of the female, displaying his nuptial attire. He raises the toothed crest on his back and slaps his tail while producing secretions from his cloacal glands.

3 CONNECTION
The male deposits his packet of sperm and then gently guides the female toward it, pushing her with his side. The female gathers the packet into her cloaca.

HABITAT	Northern Hemisphere
NUMBER OF SPECIES	360
ORDER	Urodela

NEWTS

Newt Species

Amphibians are divided into three groups, distinguished by their tails and legs. Newts and salamanders have tails and belong to the order Urodela. Some produce toxic substances for defense from predators. They are very small; the largest newt may reach 6 inches (15 cm) in length.

GREAT CRESTED NEWT
Triturus cristatus spends from three to five months of the year in the water.

Males have a crest, and females have only a yellow stripe along their backs.

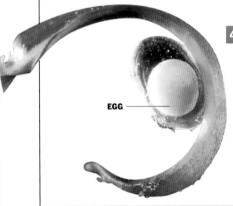

4 EGG LAYING
After the eggs are fertilized, the female finds a place to deposit them, attaching them to underwater vegetation or rocks.

EGG

EASTERN NEWT
Notophthalmus viridescens
The larvae go through a special juvenile stage called the "red phase."

DEFENSE

Some newts are highly dangerous because they release a toxic substance when attacked. One such species is the California newt. It can be recognized by its bright coloring, which serves as a warning to predators.

Anatomy of a Newt

Newts, unlike salamanders, have no grooves along their sides. Adults have elongated bodies 3–4 inches (8–10 cm) long, with well-developed tails. They have four limbs, with four toes on each front foot and four or five on each hind foot. Another peculiarity is that they have teeth in both upper and lower jaws. Their heads and eyes are relatively small. Smell is their most important sense for finding food and for social interaction.

PALMATE NEWT
Triturus helveticus
3½ inches (9 cm)
long, with a pale belly

TAIL
Newts keep
their tails as
adults.

Feeding

Like salamanders, these tiny animals are usually active at night. The smallest newts feed on small invertebrates, whereas larger newts can eat fish, amphibians, and eggs.

HIND FEET
The hind feet are
webbed in males
but not in females.

BELLY
A white or pale
belly is one of the
distinctive traits
of this species.

NEWTS AND WATER

As semiaquatic creatures, newts return to the water during mating season. They are found in North America, Europe, all of continental Asia, and Japan. Adapted to various habitats, they climb trees and dig in the ground in addition to living in the water.

**MALE'S
CREST**

MARBLED NEWT
Triturus marmoratus
spends its whole life in the
water, both as a juvenile
and as an adult.

SMOOTH NEWT
Triturus vulgaris
One of the most
colorful

5 INVERTEBRATES

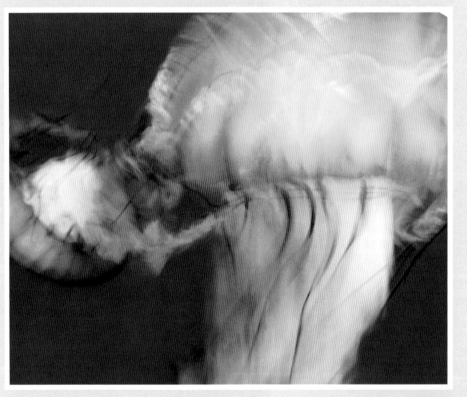

214

224

The Simplest Life-Forms

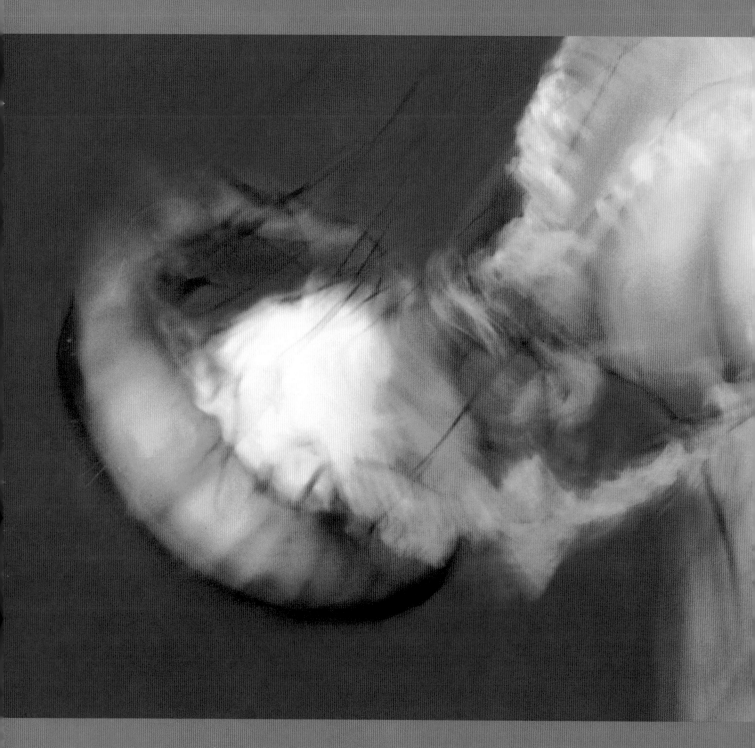

Even though some organisms, such as sponges, jellyfish, and sea anemones, look like vegetables, they belong to the animal kingdom. Many of these simple invertebrates are unable to move from one place to another; some even lack certain tissues or an entire respiratory or digestive system. Other, more developed species,

THE SIMPLICITY OF THE JELLYFISH
The jellyfish is a very simple animal with
a gelatinous consistency and with no
respiratory, digestive, or excretory systems.
It drifts around in warm ocean waters.

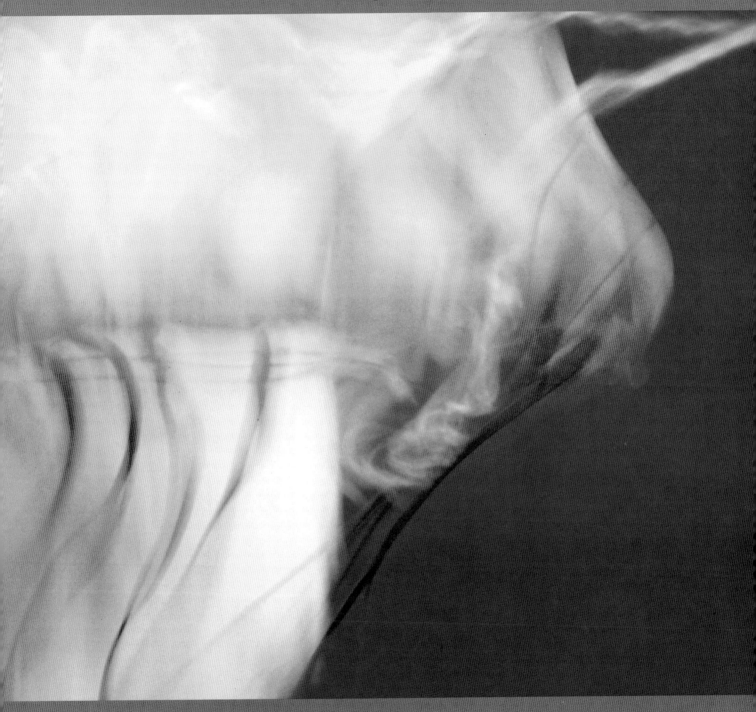

such as squid and octopus, can move about
and have become skilled marine predators.
Cephalopods are the most highly evolved
mollusks. Their heads have highly developed

eyes, a mouth with two hornlike jaws, and
tentacles with suckers to trap their prey. Some
cephalopods live in deep-sea waters, whereas
others stay close to shore.

Radial Symmetry

Many of the numerous invertebrates on the earth live in the ocean. Some, such as polyps and jellyfish, have radial symmetry—that is, their bodies are structured around an axis. A typical echinoderm, such as the starfish, has tiny, flexible, tube-shaped legs arranged like the spokes of a wheel. The animal uses them to hold onto surfaces and to move. Sponges, on the other hand, are very simple, multiple-celled animals, with many small pores that they use to feed.

RADIAL SYMMETRY

The body parts are organized around a central axis like the spokes on the wheel of a bicycle. Any plane passing through the body will divide it into two halves, each mirroring the other.

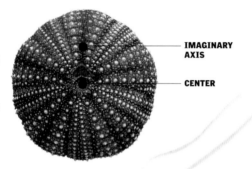

IMAGINARY AXIS

CENTER

SEA URCHIN
Strongylocentrotus franciscanus

Echinoderms

This phylum includes sea lilies, sea cucumbers, urchins, and starfish. The echinoderms have an internal skeleton made of calcified plates and a locomotion system made up of ambulacral grooves with rows of tube feet. In most echinoderm species, the endoskeleton is made of tiny calcareous plates held together by skin and muscle tissue.

ECHINODERM
means that this animal's body is covered by a spiny skin.

ECHINODERM CLASSES

ECHINOIDEA SEA URCHINS

ASTEROIDEA STARFISH

OPHIUROIDEA BRITTLE STARS

CRINOIDEA SEA LILIES

HOLOTHUROIDEA SEA CUCUMBERS

THERE ARE APPROXIMATELY

7,000

LIVING SPECIES AND 13,000 EXTINCT SPECIES OF ECHINODERMS.

Cnidarians

Cnidarians are a group of aquatic animals that includes jellyfish, hydras, sea anemones, and corals. Their cells are organized in true tissues. They have specialized cells called cnidoblasts for stinging prey and for defense. Two basic types of cnidarians are polyps and jellyfish.

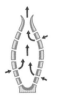

JELLYFISH
Pelagia noctiluca

CLASSIFICATION

HYDROZOA: ASEXUAL POLYP

ANTHOZOA: SEA ANEMONES AND CORALS

SCYPHOZOA: JELLYFISH

Mesoglea
Gastrovascular cavity
Mouth
Gastrodermis
Epidermis

STINGING CELL
Used for defense

1 INTACT
Cnidocyst
Nucleus
Operculum
Cnidocilium

2 DISCHARGING
Rolled-up tube
Barb

3 DISCHARGED
Unfolded stinging tube

REPRODUCTION

Most Common Habitat
COASTS OF THE UNITED STATES

Young jellyfish

Adult jellyfish

6 JELLYFISH
The polyp's body grows and begins to form jellyfish, which pile up like a stack of plates.

5 POLYP
The planula larva settles at the bottom, where it attaches to a surface. There, it develops a mouth and tentacles, and transforms into a polyp.

1 GAMETES
Adult jellyfish produce sperm and egg cells during meiosis and then release them.

2 FERTILIZATION
Fertilization takes place in the waters near the jellyfish, resulting in a zygote.

3 BLASTULA
The zygote, after a series of cell divisions, becomes a blastula, or hollow sphere, of cells.

4 PLANULA
The blastula lengthens and becomes a ciliated larva called a planula.

THERE ARE APPROXIMATELY

9,000

SPECIES OF CNIDARIANS (COELENTERATES).

Porifera

Are sessile aquatic animals. Most live at the bottom of the ocean, although there are some freshwater species. They are the simplest animals, lacking organs or true tissues, and their cells are independent to a certain extent. They are basically water-filtering bodies formed by one or more cavities. Most porifera have no definite shape, but some have radial symmetry.

THERE ARE APPROXIMATELY

5,000

SPECIES (150 ARE FRESHWATER, AND THE REST ARE MARINE).

WATER COMING OUT

Oscula
Epithelial cell
Spicule

THE WATER WITH FOOD PARTICLES ENTERS THROUGH THE POROCYTES.

Nucleus
Flagellum

TYPES OF PORIFERA ACCORDING TO ORGANIZATION

→ Direction of water flow

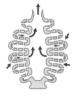

ASCON

SYCON

LEUCON

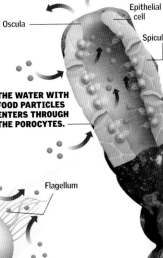

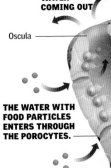

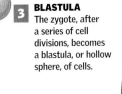

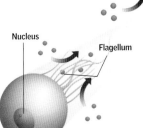

Sea Carnival

Corals and anemones, together with jellyfish, make up the phylum Cnidaria. Some characteristics they share are their bright colors, tentacles that secrete stinging substances, and a digestive system with a common opening for ingestion and excretion—the simplest digestive system in the animal kingdom. All of these organisms are quite simple. Corals generally form colonies, large groups of small polyps that stay practically immobile and feed on microorganisms brought to them by water currents. Sea anemones, on the other hand, are solitary and can trap prey despite their limited locomotion.

CORAL POLYP

Coral Reefs

Corals are small polyps with tentacles that grow at their base throughout their life, generating a calcareous exoskeleton. This skeleton forms masses, or branches. Most corals grow in colonies; the skeletons form huge calcareous masses called reefs. Corals live mostly in warm, shallow ocean waters. Their reproduction can be both sexual and asexual, by division or by gemmation. They feed on plankton.

HARD CORALS

grow over the surface of the lime-bearing substrate.

SOFT CORALS

branch out; their skeleton is not lime-based but hornlike and flexible.

CORAL WALLS

Even though some coral walls live alone, most form colonies that can grow upward at up to 3 feet (1 m) every year.

100 feet
(30 m)
THE MOST COMMON DEPTH AT WHICH CORALS GROW

TENTACLES
with stinging cells

MOUTH
Through here the animal ingests its food and excretes wastes.

HARD SKELETON
A mass that grows by the accumulation of dead polyps

LIVE TISSUE

CONNECTING TISSUE
Connects one polyp with another

GASTRIC CAVITY
Divided into several cavities in hydropolyps

CALCIUM CARBONATE

SEA ANEMONE
Any vertical plane passing through its center divides it into two equal parts.

Beautiful but Deadly

Beautiful for their shapes and colors that vary even within the same species, and dangerous for the poison they use to sting both victims and predators, sea anemones live in almost all marine latitudes, and at varying depths. Tropical marine anemones can measure up to 3 feet (1 m). They have a basal disk, which allows some species to attach to rocks, and others to slither, and still others to penetrate the seafloor. They trap live prey, even fish, with the many tentacles around their mouths.

9,000

THE NUMBER OF CNIDARIAN SPECIES IN THE WORLD

ADAPTATION OF SHAPE

To avoid being swept away in the current, the sea anemone retracts on sensing a water flow.

Water flow

Tentacles

Column

Base

CONTRACTION
The sea anemone reduces its size.

DISTENSION
By means of the retractor muscle

EXTENSION
When the water is calm

TENTACLES
With stinging cells, to hunt and move

CLOWNFISH
Inexplicably, the sea anemone's poison does not affect this species.

ORAL DISK

MOUTH

SEPTAL PERFORATION

INCOMPLETE MESENTERIUM

MESENTERIUM FILAMENT

RETRACTOR MUSCLE

PHARYNX

COMPLETE MESENTERIUM

GASTROVASCULAR CAVITY

BASAL DISK

Aquatic

Echinoderms (phylum Echinodermata) are one of the best-known groups of marine invertebrates. Sea urchins and starfish, despite their apparent differences, are part of the same group and share characteristics, such as five-radial symmetry. This phylum has an aquatic vascular system with many ambulacral grooves with tube feet, which it uses for locomotion, capturing prey, and breathing. In addition, it has an internal skeleton made of calcareous plates. These creatures lack a brain or eyes, so they use photoreceptors to sense movement.

Ambulacral Grooves

These structures are hollow cylinders with thick walls that straighten and move when a starfish injects water into certain vesicles in its body. The ambulacral grooves end in suckers that the animal uses to attach itself to objects, enabling it to move at surprising speed. These sensitive feet shrink if touched abruptly, hiding behind a rim of rigid spines that protect them from harm.

570 million years

THE LENGTH OF TIME ECHINODERMS HAVE BEEN IN EXISTENCE

STOMACH

ESOPHAGUS

MOUTH
Surrounds the food and breaks it down with stomach juices

SAC
Fills with water, expands, and clings to the surface

RADIAL CANAL
The water passes and circulates to the sacs.

SKIN
The underside is covered by spines.

Suction

The sac contracts and puts pressure on the ambulacral groove. The muscles tense and force the water back into the sac, causing suction between the groove and the surface with which it makes contact.

Closed valve

Sucker

Sac

Substrate

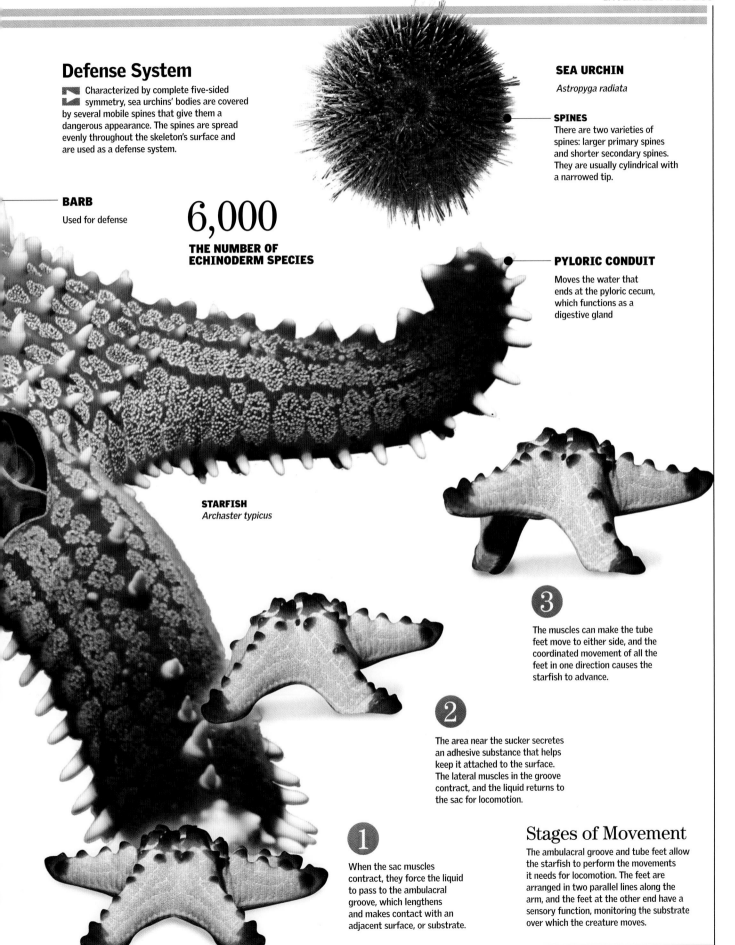

Defense System

Characterized by complete five-sided symmetry, sea urchins' bodies are covered by several mobile spines that give them a dangerous appearance. The spines are spread evenly throughout the skeleton's surface and are used as a defense system.

SEA URCHIN
Astropyga radiata

SPINES
There are two varieties of spines: larger primary spines and shorter secondary spines. They are usually cylindrical with a narrowed tip.

BARB
Used for defense

6,000
THE NUMBER OF ECHINODERM SPECIES

PYLORIC CONDUIT
Moves the water that ends at the pyloric cecum, which functions as a digestive gland

STARFISH
Archaster typicus

3
The muscles can make the tube feet move to either side, and the coordinated movement of all the feet in one direction causes the starfish to advance.

2
The area near the sucker secretes an adhesive substance that helps keep it attached to the surface. The lateral muscles in the groove contract, and the liquid returns to the sac for locomotion.

1
When the sac muscles contract, they force the liquid to pass to the ambulacral groove, which lengthens and makes contact with an adjacent surface, or substrate.

Stages of Movement

The ambulacral groove and tube feet allow the starfish to perform the movements it needs for locomotion. The feet are arranged in two parallel lines along the arm, and the feet at the other end have a sensory function, monitoring the substrate over which the creature moves.

Legless

Worms are invertebrates with long, soft bodies and no legs. They are classified into three phyla. Flatworms are the simplest type; most are parasites, although some are free-living. Nematodes have a cylindrical body with a hard outer surface. Segmented worms are more complex; they include leeches, earthworms, and sea worms. Many species have an impact on plants, animals, and humans.

Classes of Worms

Flat body

PLATYHELMINTHES

Round body

NEMATODA

Segmented body

ANNELIDA

Movement Guided by Light

Flatworms have eyespots, or light-sensitive eyes, on the front end of their bodies. When exposed to excessive light, the eyes withdraw and remain immobile.

Digestive System

In annelid worms, the digestive system extends in a straight line from the oral opening to the anus. It includes the mouth, muscular pharynx, esophagus, crop, gizzard, and intestine.

Hearts

Clitellum

Mouth

Pharynx

Reproductive system

Intestine

LOCOMOTION

Snakelike undulations along the dorsal-ventral plane

SEGMENTS

SETAE
Setae are bristlelike structures.

EARTHWORM
Lumbricus terrestris

Anus

WORM
Nematode enoplida

EPIDERMIS

PROBOSCIS
Partly folded inward

HOOKS
Hold the worm in place

28 feet
(8.5 m)
LENGTH OF THE LONGEST WORM: *PLACENTONEMA GIGANTISSIMUM*

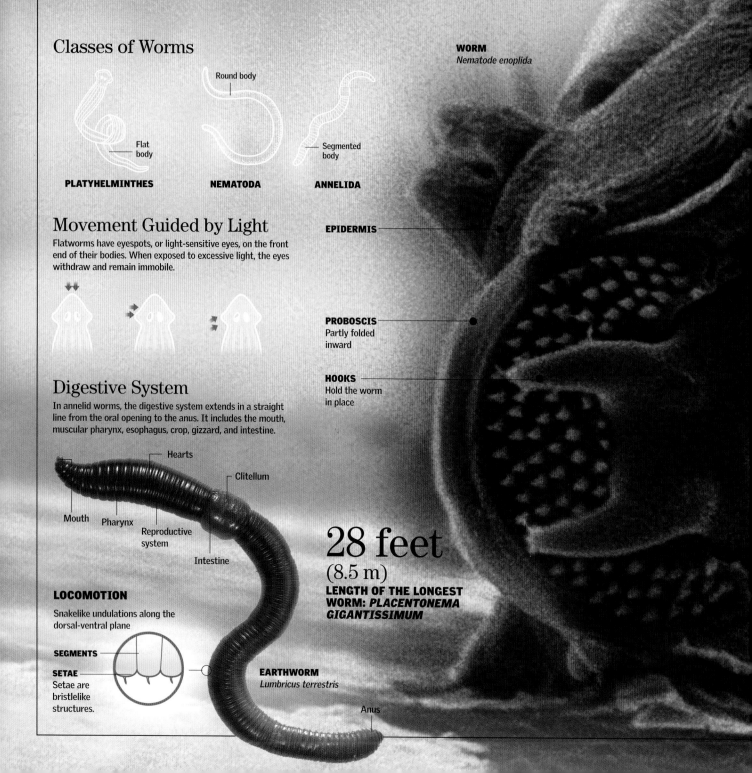

Tissues

are formed in layers and are based on the presence of internal cavities. This annelid has three layers and one cavity, the coelom, which carries fluids through the body like a hydraulic skeleton.

ECTODERM

COELOM

MESODERM

ENDODERM

DIGESTIVE CAVITY

FOOD
Bacteria and organic wastes

AT LEAST

100,000

KNOWN WORM SPECIES

NECK
Retracts and remains hidden

TISSUE
Fibrous and elastic

LEMNISCI FOOD STORAGE

Reproduction

Flatworms and annelids are usually hermaphrodites; nematodes usually have separate sexes. In some cases the worm splits into two, resulting in two new worms.

SPINES
Pierce the wall of the host

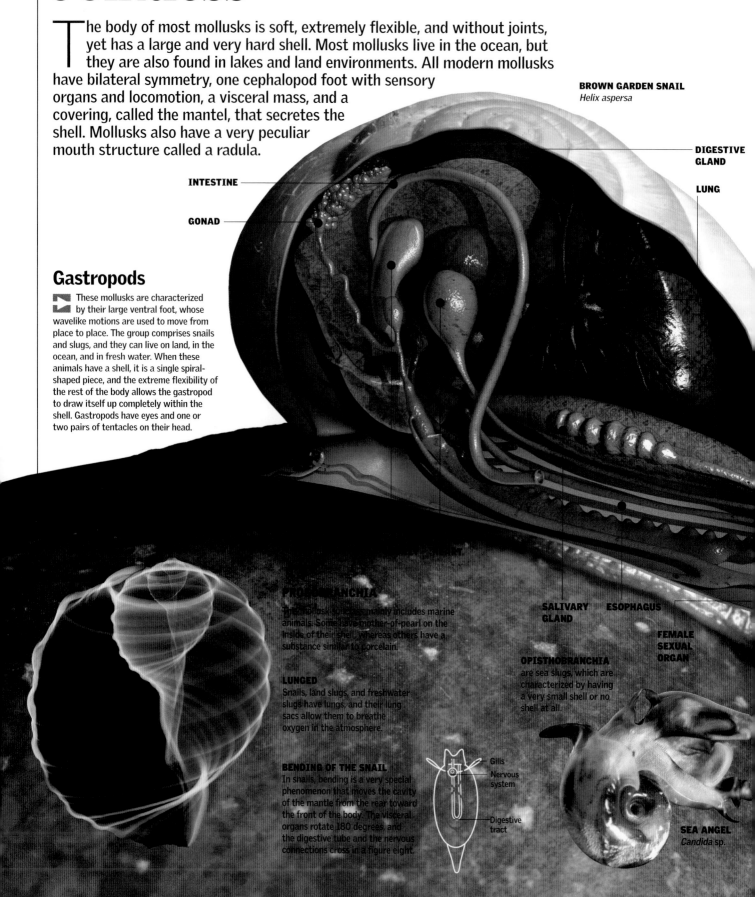

Jointless

The body of most mollusks is soft, extremely flexible, and without joints, yet has a large and very hard shell. Most mollusks live in the ocean, but they are also found in lakes and land environments. All modern mollusks have bilateral symmetry, one cephalopod foot with sensory organs and locomotion, a visceral mass, and a covering, called the mantle, that secretes the shell. Mollusks also have a very peculiar mouth structure called a radula.

BROWN GARDEN SNAIL
Helix aspersa

INTESTINE

GONAD

DIGESTIVE GLAND

LUNG

Gastropods

These mollusks are characterized by their large ventral foot, whose wavelike motions are used to move from place to place. The group comprises snails and slugs, and they can live on land, in the ocean, and in fresh water. When these animals have a shell, it is a single spiral-shaped piece, and the extreme flexibility of the rest of the body allows the gastropod to draw itself up completely within the shell. Gastropods have eyes and one or two pairs of tentacles on their head.

PROSOBRANCHIA
This mollusk subclass mainly includes marine animals. Some have mother-of-pearl on the inside of their shell, whereas others have a substance similar to porcelain.

LUNGED
Snails, land slugs, and freshwater slugs have lungs, and their lung sacs allow them to breathe oxygen in the atmosphere.

BENDING OF THE SNAIL
In snails, bending is a very special phenomenon that moves the cavity of the mantle from the rear toward the front of the body. The visceral organs rotate 180 degrees, and the digestive tube and the nervous connections cross in a figure eight.

SALIVARY GLAND   **ESOPHAGUS**

FEMALE SEXUAL ORGAN

OPISTHOBRANCHIA
are sea slugs, which are characterized by having a very small shell or no shell at all.

Gills
Nervous system
Digestive tract

SEA ANGEL
Candida sp.

Bivalves

Mollusks with a shell divided into two halves. The two parts of the shell are joined by an elastic ligament that opens the shell, abductor muscles that close the shell, and the umbo, a system of ridges that helps the shell shut together. Almost all bivalves feed on microorganisms. Some bury themselves in the wet sand, digging small tunnels that let in water and food. The tunnels can be from a fraction of an inch long to over a yard long.

SCALLOP
Pecten jacobaeus

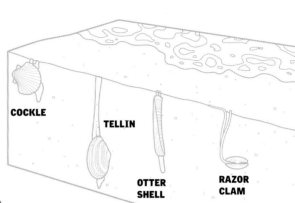

COCKLE

TELLIN

OTTER
SHELL

RAZOR
CLAM

LAMELLIBRANCHIATA

include most bivalves. They use gills to breathe and to feed. They have no differentiated head, eyes, or extremities. They can grow up to 5 inches (13 cm) long, and they rest on the ocean floor.

GREEN MUSSEL
Perna viridis

Under the Sand

Many mollusks live buried under the sand in order to hide from predators and the effects of waves, wind, and sudden changes in temperature.

PROTOBRANCHIA

This class includes bivalves with a split lower foot, called a sole. Bivalves use their gills only to breathe. This subclass includes small bivalves 1/2 inch (13 mm) wide, called nutclams (*Nucula nitidosa*).

100,000

**THE NUMBER OF LIVING
MOLLUSK SPECIES; AS MANY
MORE HAVE BECOME EXTINCT**

Cephalopods

Cuttlefish, octopus, squid, and nautilus are called cephalopods because their extremities, or tentacles, are attached directly to their heads. These predators are adapted to life in the oceans, and they have quite complex nervous, sensory, and motion systems. Their tentacles surround their mouths, which have a radula and a powerful beak. Cephalopods can be 1/3 inch (1 cm) long to several yards long.

RADULA

NAUTILUS
Nautilus sp.

COLEOIDEA

Cephalopods of this class have a very small internal shell, or none at all, and only two gills. Except for the nautilus, this class includes all cephalopods alive today—octopus, cuttlefish, and squid.

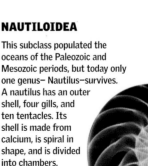

COMMON
CUTTLEFISH
...officinalis

NAUTILOIDEA

This subclass populated the oceans of the Paleozoic and Mesozoic periods, but today only one genus– Nautilus–survives. A nautilus has an outer shell, four gills, and ten tentacles. Its shell is made from calcium, is spiral in shape, and is divided into chambers.

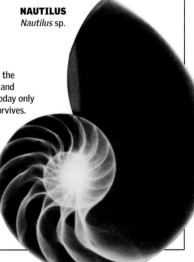

Generating Added Value

Bivalves are sought after and cultivated for their pearls. Pearls are said to be the queen of gems, because they were discovered more than 4,000 years ago and were used as important symbols in many ancient cultures. In spite of their high price, pearls start out as a nuisance for the animal that creates them, which could be an oyster, a clam, or a mussel. Oysters produce the most valuable pearls, which are noted for their luster.

Pearl Formation

Occasionally, grains of sand or parasites accidentally become lodged in an oyster's body and cannot escape. To relieve the annoyance, the oyster begins a defensive action and secretes a smooth, hard, crystalline substance called nacre around the object. Cultured pearls are formed from particles that are intentionally inserted into an oyster.

SHELL
Composed of two pieces or valves

INNER SURFACE OF THE SHELL
Sensory tentacles enable the oyster to detect light and darkness.

DIGESTIVE GLAND
Its cells absorb and digest food particles.

1

INCUBATION

Pearl cultivation began in Japan. It consists of inserting into the body of a live oyster a small, round particle made from the shell of a freshwater bivalve. The oyster secretes mother-of-pearl substances from a gland in its liver to cover the object, and the pearl begins to grow.

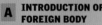 **A INTRODUCTION OF FOREIGN BODY**

GRAIN OF SAND

B The oyster secretes nacre to cover it.

TONGUE

2

GROWTH OF THE PEARL

New, uniform layers are constantly added to the pearl, and the cultivator leaves the pearl in place until it reaches the required diameter and quality. During the process, humans intervene only to provide the oysters in farms with the right temperature, water currents, and cleanliness to favor the growth of pearls.

3 to 8 years

TIME IT TAKES FOR A PEARL TO GROW

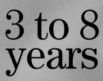

ORGANIC LAYER

ARAGONITE CRYSTAL

LAYERS OF NACRE ON THE PEARL

LAYERS OF NACRE ON THE SHELL

OYSTERS

TIED WITH ROPES

HANGING OYSTERS

These oysters are suspended from bamboo rafts in areas with abundant plankton.

 3

HARVEST

Cultured pearls make up 95 percent of the pearls currently sold. About 500 million pearls are produced every year. However, pearl farming is a demanding and difficult business because of the nature of pearl oysters: out of 100 oysters cultivated, only 30 will be harvested.

Types of Pearls

They can be round or elongated like a grain of rice.

NATURAL PEARL

CULTURED PEARLS

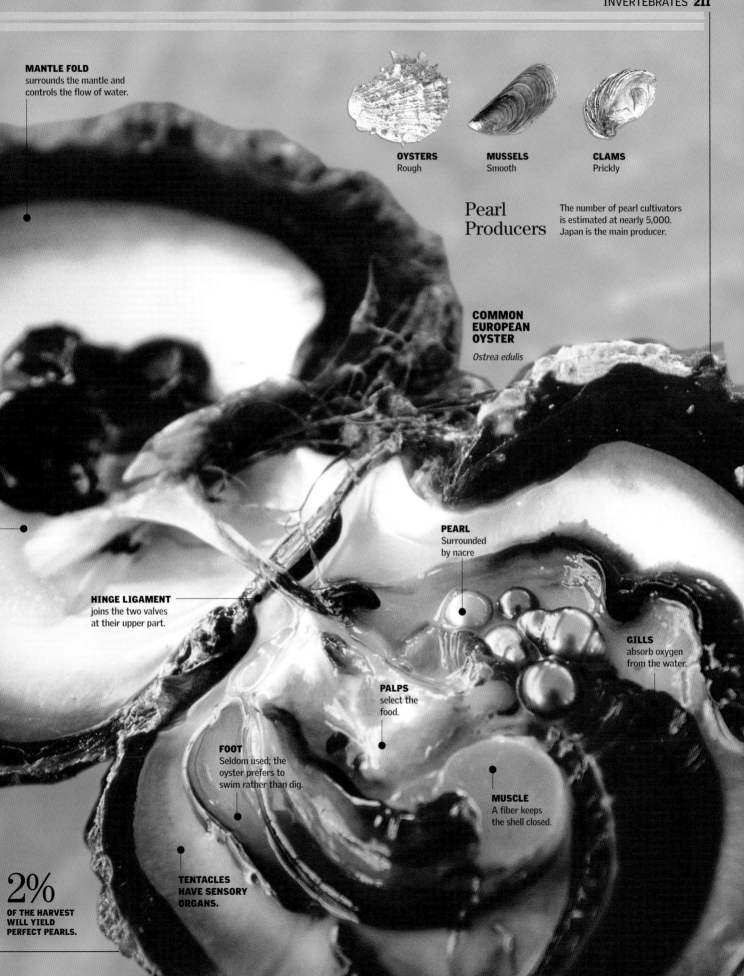

MANTLE FOLD
surrounds the mantle and
controls the flow of water.

OYSTERS
Rough

MUSSELS
Smooth

CLAMS
Prickly

Pearl Producers

The number of pearl cultivators
is estimated at nearly 5,000.
Japan is the main producer.

**COMMON
EUROPEAN
OYSTER**

Ostrea edulis

PEARL
Surrounded
by nacre

HINGE LIGAMENT
joins the two valves
at their upper part.

GILLS
absorb oxygen
from the water.

PALPS
select the
food.

FOOT
Seldom used; the
oyster prefers to
swim rather than dig.

MUSCLE
A fiber keeps
the shell closed.

**TENTACLES
HAVE SENSORY
ORGANS.**

2%
**OF THE HARVEST
WILL YIELD
PERFECT PEARLS.**

Powerful Tentacles

The eight-tentacled octopus is one of the few large ocean cephalopods to live in deep water. It is usually found on the rocky or sandy bottoms of shallow waters near the mouths of rivers. It generally moves slowly, sometimes moving in brief spurts, but it can reach great speeds when hunting or fleeing. Some are quite intelligent, having highly evolved brains.

Masters of Color

For the octopus, taking on the color of the ocean floor is a camouflage strategy to hide from its prey. In deeper waters, another tactic is to become luminescent to attract the prey. But when the octopus changes colors while doing a certain dance, it is trying to attract the opposite sex.

HEAD
The head compresses and expands, depending on the octopus's breathing and movements. The head contains the brain but without a rigid protective structure.

EYES
Are located on the head. The octopus's sense of sight is exceptionally well developed.

SKIN
The skin is a highly elastic membrane that completely covers the octopus.

Attack

To attack, the octopus points its funnel in the direction opposite to its motion. The common octopus (*Octopus vulgaris*), a species that can grow up to 40 inches (1 m) long and inhabits the Mediterranean Sea and North Atlantic Ocean, moves among the rocks on the seafloor, preferably at night. It surprises its prey and makes skillful use of its tentacles and jaws, which can rotate.

1 The funnel muscles can act as a mechanism for fleeing. Rather than directing the funnel forward, though, the octopus directs it to advance toward its prey.

2 The tentacles stretch forward and outward as the octopus advances.

3 Using the wide area at the base of its tentacles, it envelops the prey.

Large Predators

Depending on its size, an octopus (like other large cephalopods, such as the nautilus, cuttlefish, and squid) is carnivorous and eats both fish and other invertebrates: mollusks and crustaceans, especially crabs. It secretes a venom with its saliva to finish killing the prey before swallowing it.

Speedy Escape

The flow of water into and out of the funnel is regulated by alternately contracting and relaxing ring-shaped muscles and long muscles. By regulating the force at which the water is expelled, the octopus can flee at high speed through a kind of jet propulsion. The octopus moves in the direction its head is pointing, with its tentacles outstretched.

The ring-shaped muscles relax, and the long muscles contract. Water enters.

When the ring-shaped muscles contract, they expel a jet of water that propels the octopus backward.

SELF-DEFENSE WITH INK
A gland located near the anus contracts when the octopus senses danger, expelling a fluid that creates a dark cloud in the water.

4 miles per hour (6 km/h)

MAXIMUM SPEED OF A FLEEING OCTOPUS. ITS SPEED IS COMPARABLE TO THAT OF A FAST-WALKING HUMAN.

1 BREATHING

Head

H_2O

2 PROPULSION

Funnel

Gill

The Funnel

The funnel is the exit from the octopus's respiratory cavity. It is also extremely important for the creature's movement. The gills, inside the mantle, absorb oxygen from the water. When the cavity fills, the gills exchange oxygen for carbon dioxide to be emptied from the cavity.

TENTACLES
All eight tentacles have the same length. In the male, one tentacle functions as a genital organ.

MUSCLES
Powerful and versatile, with self-controlled movements, an octopus can move the entire weight of its body.

SUCKERS
Arranged in two rows on the lower surfaces for clinging to rocks and for grasping prey.

Grasping Ability

An octopus often crawls among the rocks. Using the system of suckers, or adhesive disks, on its tentacles, an octopus clings to the seafloor or supports itself by attaching the suckers to the surfaces it encounters. By grasping with its forward tentacles, it can drag the rest of its body in that direction.

1 RELAXED MUSCLE

chitinous ring

SUCTION

2 CONTRACTED MUSCLE

Crustaceans and Arachnids

S piders, snakes, ticks, and mites all belong to the same class, Arachnida. They are covered with sensory hairs so tiny that they cannot be seen by the naked eye.

In Greek mythology, Arachne was a woman who challenged the goddess Athena to weave faster than she herself could. This angered the goddess, who turned Arachne into a spider, forcing her to weave

COLORFUL ARACHNID
Some species of the family Trombidiidae catch our attention because of their intense red color and the velvetlike appearance of their hair.

forever. That is where these creatures get their name. Within the world of crustaceans, well-known animals, such as the shrimp, lobster, and crab, are also discussed in this chapter. You will find details about their anatomy, their differences and similarities, and the way in which they live that will surprise you. Some species breathe through gills and also breathe through their skin.

Colorful Armor

Even though they inhabit all known environments, crustaceans are most closely identified with the aquatic environment. That environment is where they were transformed into arthropods with the most evolutionary success. Their bodies are divided into three parts: the cephalothorax, with antennae and strong mandibles; the abdomen, or pleon; and the back (telson). Some crustaceans are very small: sea lice, for instance, are no larger than $1/100$ inch (0.25 mm). The Japanese spider crab, on the other hand, is more than 9 feet (3 m) long with outstretched legs, because it has legs in both the abdomen and the thorax in addition to two pairs of antennae.

WOOD LOUSE

(Armadillidium vulgare)

This invertebrate, belonging to the order Isopoda, is one of the few terrestrial crustaceans, and it is probably the one best adapted to life outside the water. When it feels threatened, it rolls itself up, leaving only its exoskeleton exposed. Even though it can reproduce and develop away from the water, it breathes through gills. The gills are found in its abdominal appendages and for this reason must be kept at specific humidity levels. That is also why the wood louse seeks dark and humid environments, such as under rocks, on dead or fallen leaves, and in fallen tree trunks.

Extended animal

EXOSKELETON
Divided into independent parts

Malacostraca

is the name given to the class of crustaceans that groups crabs together with sea lobsters, shrimp, wood lice, and sea lice. The term comes from Greek, and it means "soft-shelled." Sea and river crabs have ten legs, and one pair of these legs is modified in a pincer form. Malacostraca are omnivorous and have adapted to a great variety of environments; the number of segments of their exoskeleton can vary from a minimum of 16 to more than 60.

APPENDAGES
consist of a lower region from which two segmented branches grow, one internal (endopod) and the other external (exopod).

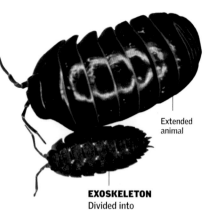

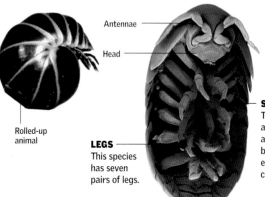

Antennae

Head

Rolled-up animal

LEGS
This species has seven pairs of legs.

SEGMENTS
The back segments are smaller, and when they bend, they help enclose the animal completely.

Anus

THE PACIFIC SPIDER CRAB CAN WEIGH UP TO

45 pounds
(20 kg).

BARNACLES WITHOUT A SHELL

BARNACLE COLONY

Together Forever

At birth, barnacles (*Pollicipes cornucopia*) are microscopic larvae that travel through the sea until they reach a rocky coast. Then they attach themselves to the shore by means of a stalk, which they develop by the modification of their antennae, and then form a shell. Once they are attached, they remain in one spot for the rest of their lives, absorbing food from the water. Barnacles are edible.

BARNACLE TRANSVERSAL CUT

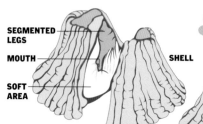

SEGMENTED LEGS

MOUTH

SOFT AREA

SHELL

LEGS EXTENDED TO CATCH FOOD

SHELL

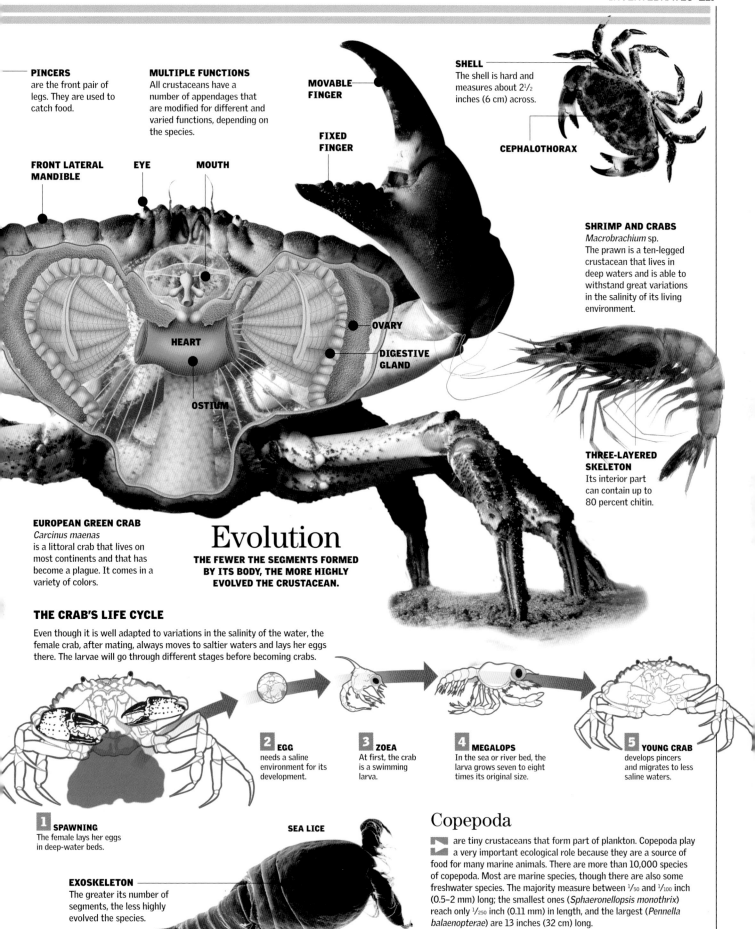

PINCERS
are the front pair of
legs. They are used to
catch food.

MULTIPLE FUNCTIONS
All crustaceans have a
number of appendages that
are modified for different and
varied functions, depending on
the species.

**MOVABLE
FINGER**

**FIXED
FINGER**

SHELL
The shell is hard and
measures about 2¹/₂
inches (6 cm) across.

CEPHALOTHORAX

**FRONT LATERAL
MANDIBLE**

EYE

MOUTH

HEART

OVARY

**DIGESTIVE
GLAND**

OSTIUM

SHRIMP AND CRABS
Macrobrachium sp.
The prawn is a ten-legged
crustacean that lives in
deep waters and is able to
withstand great variations
in the salinity of its living
environment.

**THREE-LAYERED
SKELETON**
Its interior part
can contain up to
80 percent chitin.

EUROPEAN GREEN CRAB
Carcinus maenas
is a littoral crab that lives on
most continents and that has
become a plague. It comes in a
variety of colors.

Evolution
**THE FEWER THE SEGMENTS FORMED
BY ITS BODY, THE MORE HIGHLY
EVOLVED THE CRUSTACEAN.**

THE CRAB'S LIFE CYCLE

Even though it is well adapted to variations in the salinity of the water, the
female crab, after mating, always moves to saltier waters and lays her eggs
there. The larvae will go through different stages before becoming crabs.

1 SPAWNING
The female lays her eggs
in deep-water beds.

2 EGG
needs a saline
environment for its
development.

3 ZOEA
At first, the crab
is a swimming
larva.

4 MEGALOPS
In the sea or river bed, the
larva grows seven to eight
times its original size.

5 YOUNG CRAB
develops pincers
and migrates to less
saline waters.

SEA LICE

EXOSKELETON
The greater its number of
segments, the less highly
evolved the species.

Copepoda

are tiny crustaceans that form part of plankton. Copepoda play
a very important ecological role because they are a source of
food for many marine animals. There are more than 10,000 species
of copepoda. Most are marine species, though there are also some
freshwater species. The majority measure between ¹/₅₀ and ¹/₁₀₀ inch
(0.5–2 mm) long; the smallest ones (*Sphaeronellopsis monothrix*)
reach only ¹/₂₅₀ inch (0.11 mm) in length, and the largest (*Pennella
balaenopterae*) are 13 inches (32 cm) long.

Sharp Front Legs

Crustaceans have appendages that generally branch in two directions and are adapted to aquatic life. A characteristic shared by all crustaceans is their articulated shell, which leaves two pairs of antennae uncovered. They also have a pair of mandibles, two pairs of maxillae, and a pair of appendages in each segment of the body. Their pincers have enough strength so they can trap their prey and feed themselves. The class Malacostraca includes lobsters, crabs, shrimp, and prawns, among other animals.

Shrimp

is the name for about 2,000 species of crustaceans of the suborder Natantia. Shrimp are characterized by their semitransparent and flat bodies, with appendages modified for swimming, and by their long antennae. Their length varies between 1/10 inch and about 8 inches (from a few mm to 20 cm), depending on the species. They live in salt water, brackish water, and fresh water. They survive by burying themselves for almost the entire day and coming out at dusk to catch their food.

55,000

LIVING SPECIES AND AS MANY FOSSIL SPECIES ARE PART OF THIS GROUP OF INVERTEBRATES.

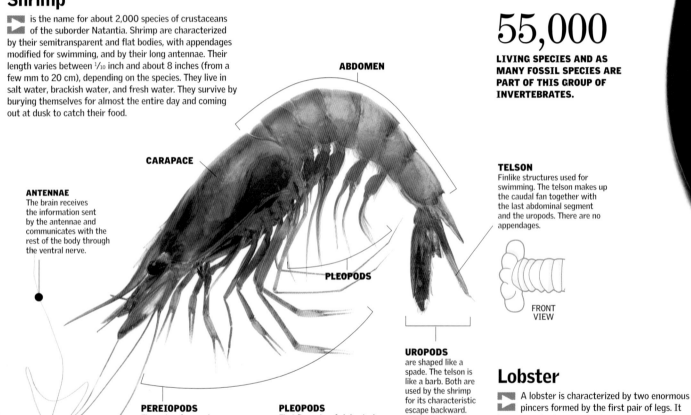

ABDOMEN

CARAPACE

ANTENNAE
The brain receives the information sent by the antennae and communicates with the rest of the body through the ventral nerve.

TELSON
Finlike structures used for swimming. The telson makes up the caudal fan together with the last abdominal segment and the uropods. There are no appendages.

PLEOPODS

FRONT VIEW

UROPODS
are shaped like a spade. The telson is like a barb. Both are used by the shrimp for its characteristic escape backward.

SHRIMP
Caridea

PEREIOPODS
Five pairs of appendages

FIRST THREE PAIRS
are used to feed itself. The pincers catch and hold prey.

LAST TWO PAIRS
work as walking legs that are aided by the pleopods.

PLEOPODS
First five pairs of abdominal appendages

FIRST TWO PAIRS
have been adapted for sexual functions.

LAST THREE PAIRS
are similar to each other and are used to swim.

Lobster

A lobster is characterized by two enormous pincers formed by the first pair of legs. It lives on rocky bottoms in shallow water, and it migrates seasonally toward the coast in summer and to greater depths in winter. The lobster is typically a nocturnal animal seeking its food when the sun sets. Its food consists mainly of mollusks, bivalves, worms, and fish.

Crab

Of all crustaceans, the crab has surprising mobility and agility. It has five pairs of legs, four of which are walking legs, despite the fact that it moves laterally instead of forward. The crablike movement is due to the placement of its legs and the general design of its body. A crab's walk is funny, but its technique is effective for both swimming and walking, even over such varied surfaces as beach sand, rock, and—for some species—tree branches.

AT REST
The body remains close to the ground, the center of gravity is lowered, and movements are slow and rhythmic.

PENDULUM SLOW WALK
The body operates like the weight of a pendulum. Close to the ground, it saves energy by moving in a swinging motion.

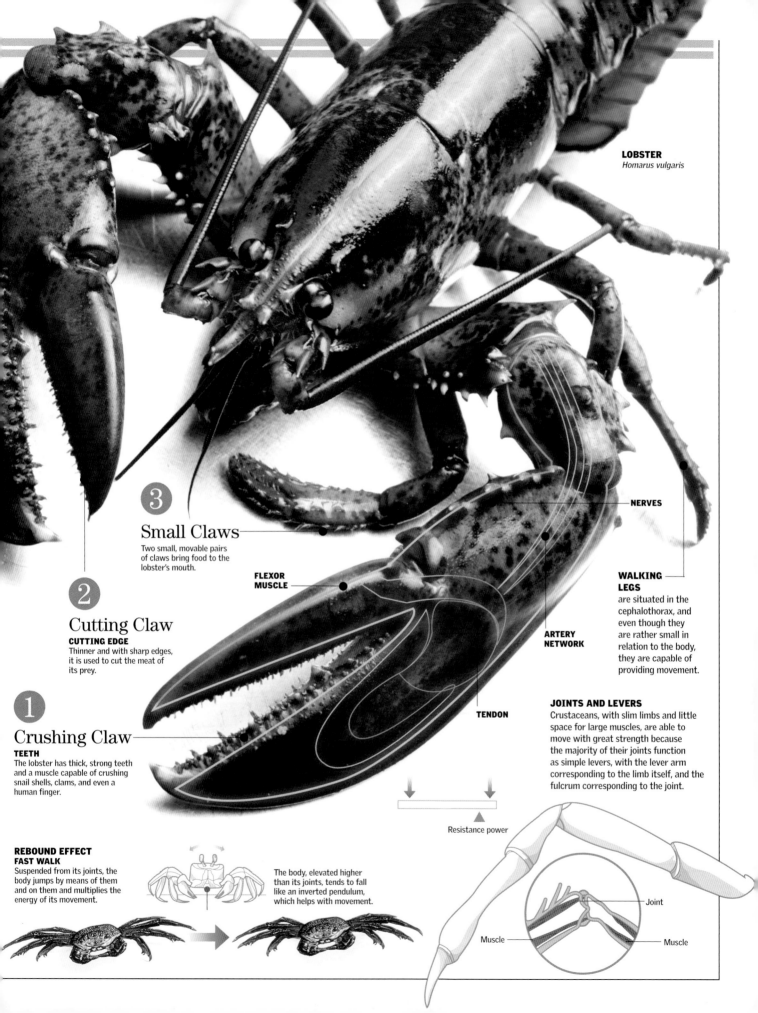

LOBSTER
Homarus vulgaris

NERVES

③

Small Claws
Two small, movable pairs
of claws bring food to the
lobster's mouth.

**FLEXOR
MUSCLE**

**WALKING
LEGS**
are situated in the
cephalothorax, and
even though they
are rather small in
relation to the body,
they are capable of
providing movement.

②

Cutting Claw
CUTTING EDGE
Thinner and with sharp edges,
it is used to cut the meat of
its prey.

**ARTERY
NETWORK**

JOINTS AND LEVERS
Crustaceans, with slim limbs and little
space for large muscles, are able to
move with great strength because
the majority of their joints function
as simple levers, with the lever arm
corresponding to the limb itself, and the
fulcrum corresponding to the joint.

①

TENDON

Crushing Claw
TEETH
The lobster has thick, strong teeth
and a muscle capable of crushing
snail shells, clams, and even a
human finger.

Resistance power

**REBOUND EFFECT
FAST WALK**
Suspended from its joints, the
body jumps by means of them
and on them and multiplies the
energy of its movement.

The body, elevated higher
than its joints, tends to fall
like an inverted pendulum,
which helps with movement.

Joint

Muscle

Muscle

In the Middle of the Chain

Zooplankton include thousands of distinct species belonging to very different groups. Among these species are protists, coelenterates, worms, crustaceans, and other small, weakly swimming organisms. Unicellular, eukaryotic protists constitute a large group of species of zooplankton. They constitute an extensive and varied community in the food network. The phytoplankton, which are capable of photosynthesis, provide food for the zooplankton. Phytoplankton also serve as food for echinoderms, crustaceans, and larval-stage fish. Once they grow up, the larvae serve as food for schools of small fish, which are, in turn, food for larger fish, including plankton-feeding whales that sometimes eat these small fish.

Malacostraca

Are typically oceanic, although some have adapted to fresh water, and others are even adapted to life on land. All have a body divided into a 13-segment cephalothorax with 13 pairs of appendages, a stomach with six segments, and, at the extreme posterior, an unsegmented telson.

KRILL
Euphausia superba
Is one of the most abundant and successful species on the earth. Krill can live for five to ten years, experiencing ten moltings before reaching their maximum length. Krill typically emit a greenish light that can be seen at night.

REAL SIZE
1½ INCHES
(3.8 CM)

EYE
Krill have only one large, compound, black eye.

LEGS
With their feathery legs, krill filter out the small algae on which they feed.

6,600 feet
(2,000 m)
DEPTH TO WHICH SWARMS OF KRILL MAY GATHER

HOW IT FLEES

The krill makes use of its telson, comprising five paddles, to drive itself through the water. It reaches great speed and moves jumping forward and backward. These crustaceans group in giant schools, with thousands of individuals concentrated in each cubic yard of water.

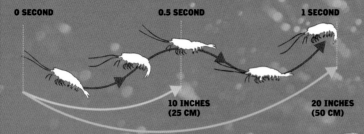

0 SECOND

0.5 SECOND

1 SECOND

10 INCHES (25 CM)

20 INCHES (50 CM)

LUMINESCENCE
Each specimen of krill has a photophore in its abdomen, a structure that allows it to emit light because of a chemical reaction that involves oxygen and various other chemical compounds called luciferin, luciferase, and adenosine triphosphate (ATP). One order of crustaceans is generically known as krill.

TROPHIC CHAIN

The food cycle is initiated with a vegetable producer, which begins the chain for consumers. Those that feed on producers are primary consumers, those that feed on primary consumers are secondary consumers, and so on.

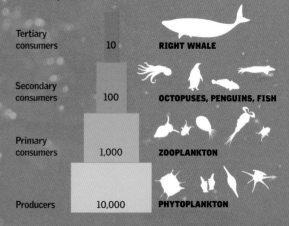

Tertiary consumers	10	RIGHT WHALE
Secondary consumers	100	OCTOPUSES, PENGUINS, FISH
Primary consumers	1,000	ZOOPLANKTON
Producers	10,000	PHYTOPLANKTON

Copepods

are aquatic microcrustaceans, although terrestrial ones also exist. They are found in fresh water as well as in salt water. They feed on phytoplankton and are an important component of plankton, which at the same time serves as food for numerous marine animals.

CYCLOPOID COPEPOD
Megacyclops viridis
Cyclopoid copepod larvae are luminescent. After their developmental stage they begin to swim freely. The cyclopoid copepod lives in fresh water. It is among the most numerous invertebrate species found in Europe.

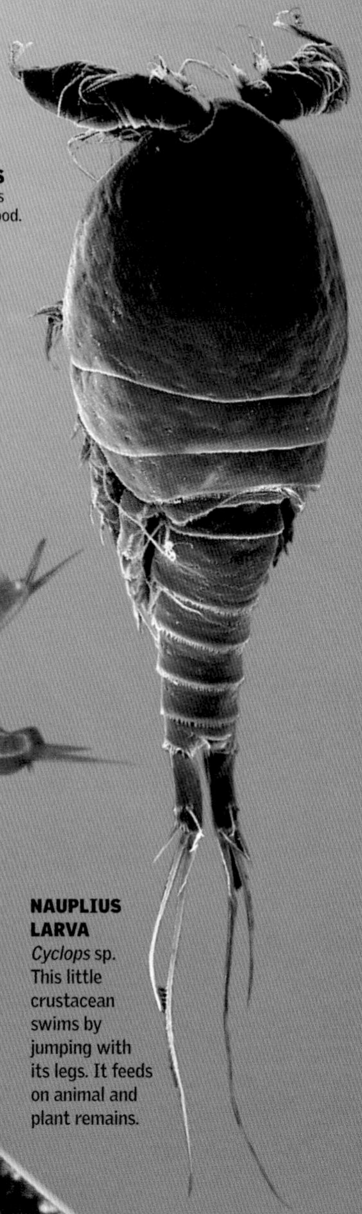

12,000
SPECIES OF COPEPODS

LARGE APPENDAGES
They form very fine combs that filter the water for food.

REAL SIZE

¹/₂₅ **INCH
(2 MM)**

NAUPLIUS LARVA
Cyclops sp. This little crustacean swims by jumping with its legs. It feeds on animal and plant remains.

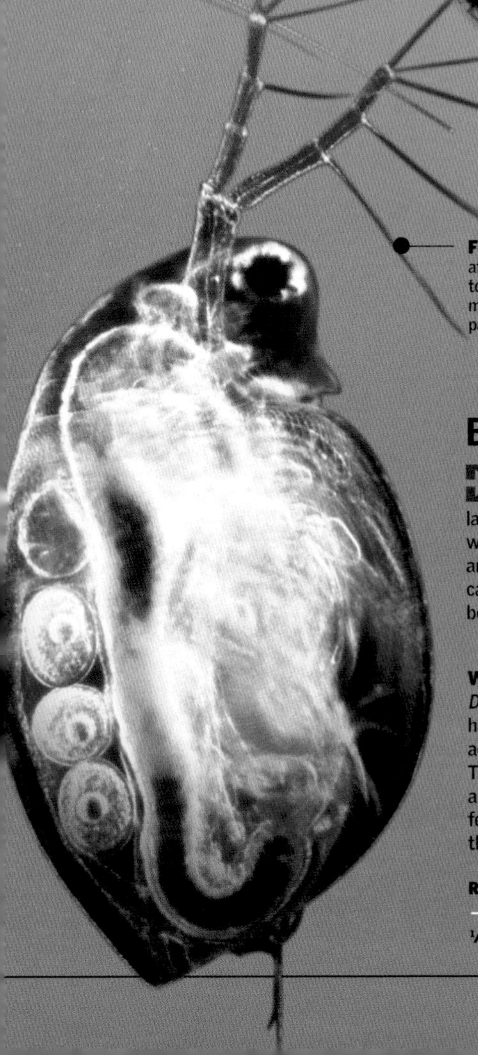

FEET
attract the water current toward the larva's mouth, which little particles can enter.

Branchiopods

are the most primitive types of crustaceans. They live in lakes and ponds throughout the world. They have compound eyes and usually a protective plate, or carapace. They also have many body segments.

WATER FLEA
Daphnia sp.
has two pairs of antennae and feet adapted to swimming and grasping. The second antenna pair serves as a locomotive organ. The water flea feeds on microscopic seaweed and the remains of dead animals.

REAL SIZE

¹/₁₀ **INCH (3 MM)**

6 to 8 weeks
IS THE AVERAGE LIFESPAN OF A WATER FLEA.

A Special Family

Arachnids make up the largest and most important class of chelicerata. Among them are spiders, scorpions, fleas, ticks, and mites. Arachnids were the first arthropods to colonize terrestrial environments. The fossil remains of scorpions are found beginning in the Silurian period, and they show that these animals have not undergone major changes in their morphology and behavior. The most well-known arachnids are the scorpions and spiders.

GIANT HOUSEHOLD SPIDER
Tegenaria duellica
This spider is distinguished by its long legs in relation to its body.

The female can transport up to 30 offspring on its back.

The claws hold the prey and immobilize it.

Scorpions

Feared by people for ages, the scorpion is characterized by the fact that its chelicerae (mouth parts that in scorpions are large) and pedipalps form a pincer. The body is covered with a chitinous exoskeleton that includes the cephalothorax and abdomen.

EMPEROR SCORPION
Pandinus imperator
Like other scorpions, it has a stinger crisscrossed by venomous glands. It measures between 5 and 7 inches (12 and 18 cm) long, although some have reached a length of 8 inches (20 cm).

PEDIPALPS
The terminal pedipalp forms a copulating organ through which the male inseminates the female.

PEDIPALPS
act as sensory organs and manipulate food. Males also use them for copulation.

CHELICERAE
move up and down. In the more primitive spiders (such as tarantulas), the chelicerae move side to side like a pincer.

SALIVA GLANDS

TICK

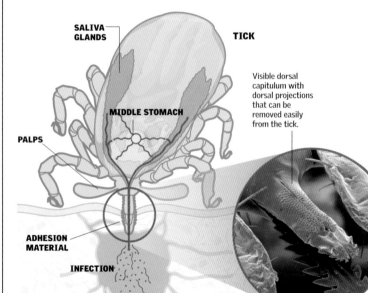

Visible dorsal capitulum with dorsal projections that can be removed easily from the tick.

MIDDLE STOMACH

PALPS

ADHESION MATERIAL

INFECTION

Mites and Ticks

Both are members of the Acari order. They are differentiated by their size. Mites are smaller; ticks may measure up to an inch in length (several centimeters). Mites have many diverse forms and are parasites of animals and plants. Ticks have a common life cycle of three stages: larva, nymph, and adult, during which they live off the blood of their hosts.

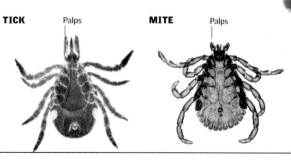

TICK   Palps   **MITE**   Palps

100,000

IS THE NUMBER OF SPECIES OF ARACHNIDS THOUGHT TO EXIST IN THE WORLD.

EXOSKELETON

Growth happens through molting, a process by which the spider gets rid of its old exoskeleton. In its youth the spider grows through successive moltings (up to four a year), and once it reaches adulthood, it goes through a yearly change.

1 The front edge of the shell comes off, and the tegument separates from the abdomen.

2 The spider raises and lowers its legs until the skin slips and falls.

3 It removes the old exoskeleton, and the new one hardens on contact with the air.

CEPHALOTHORAX (PROSOMA)

ABDOMEN (OPISTHOSOMA)

CHELICERAE

HEART

SIMPLE EYE

INTESTINE

CLOACA

VENOM GLAND

STOMACH

OVARIES

LUNG

GENITAL ORIFICE

SILK GLAND

FEMUR

PATELLA

TIBIA

WALKING LEGS

The spider has four pairs of legs for walking. The hairs help it to recognize terrain.

METATARSUS

Spiders

are the most common arthropods. They have the surprising property of secreting a substance that, on contact with the air, creates very fine threads that spiders skillfully manage for diverse purposes. Once a female spider mates, she deposits her eggs inside a cocoon of special silk, called an egg sack. The appearance of spiders is unmistakable: the two main sections of the body, the thorax (also called a prosoma) and the abdomen (also called an opisthosoma), are united by a narrow stalk (the pedicel). Spiders have four pairs of eyes, whose distinctive size and placement help characterize different families of spiders. Their chelicerae end in fangs that carry conduits from venom glands. Spiders kill their prey by using their chelicerae to apply venom.

TARSUS

WITH ITS LEGS SPREAD OUT, A SPIDER CAN MEASURE

12 inches

(30 cm) **IN LENGTH.**

Amblypygi

Small arachnids measure between 1/5 and 2 inches (4–45 mm). The chelicerae are not as large, although the pedipalps are strong and are used to capture prey. The first pair of legs are modified touch-and-sensing appendages, whereas the last three take care of movement. Because of a spider's flattened body, its walk is similar to that of the crab.

PEDIPALP ARACHNID
Phryna grossetaitai

Insects

Insects make up the largest and most varied group of arthropods. Most reproduce easily, and there are insects adapted to any environment. Their bodies are protected by a form of armor. Arthropods are currently believed to be the only living things capable of surviving a nuclear winter. They have highly developed sensory organs that enable them

A PECULIAR SENSE OF SIGHT
This tropical insect has a pair
of eyes on each side of its body,
giving it a very wide field
of vision.

to see long distances. The diversity and sheer number of insect species, estimated at 1.5 million, are a testimony to their evolutionary success. They have been successful, in part, because they are small, need less food than larger organisms, and have extraordinarily developed means of movement that keep them from being easy victims for predators.

The Secret of Success

Sensory antennae, appendages on the head that can be used to chew, crush, or grab, highly developed eyes on the sides of the head, and pairs of jointed legs with functions that depend on the species—all are outstanding common features of insects and millipedes (subphylum Myriapoda). Insects, also called hexapods, have six legs attached to the thorax. Myriapods are multi-segmented arthropods that have developed only on land.

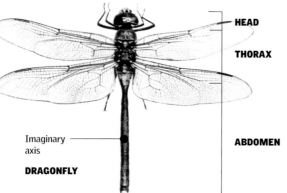

HEAD

THORAX

ABDOMEN

Imaginary axis

DRAGONFLY

BILATERAL SYMMETRY

The entire body of insects and myriapods is composed of pairs, arranged along an imaginary axis that passes from the head to the lower end of the abdomen.

Two Pairs of Wings

Some ancient species had three pairs of wings. Today, however, insects have one or two pairs. Butterflies, dragonflies, bees, and wasps use two pairs to fly, but other insects fly with only one pair.

HIND WINGS

AT REST
Dragonflies can place their wings against their bodies.

OPEN CIRCULATION

A tubular heart pumps the hemolymph (blood) forward through the dorsal aorta. Accessory contracting organs help push the blood into the wings and legs.

APPENDAGE
contains the genital organs.

SEGMENTED REGIONS

Insects' bodies are divided into three parts: the head (6 segments), the thorax (3 segments), and the abdomen (up to 11 segments).

SPIRACLES
Small entrances to the tracheae

1 million
KNOWN INSECT SPECIES

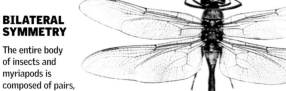

RESPIRATORY SYSTEM

Land-dwelling arthropods breathe with tracheae. Through branching tubes (tracheoles), air containing oxygen is brought directly to each cell, and carbon dioxide is eliminated.

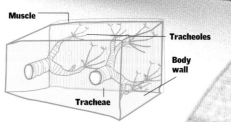

Muscle

Tracheoles

Body wall

Tracheae

Legs Adapted for Type of Use

The shape of the arthropod legs shown here is closely related to their use and to the arthropod's habitat. Some species have taste and touch receptors on their legs.

Sacs

STRUCTURE
gives the wings great stability.

LEGS

WALKING COCKROACH

JUMPING GRASSHOPPER

SWIMMING WATER SCORPION

DIGGING MOLE CRICKET

GATHERING BEE

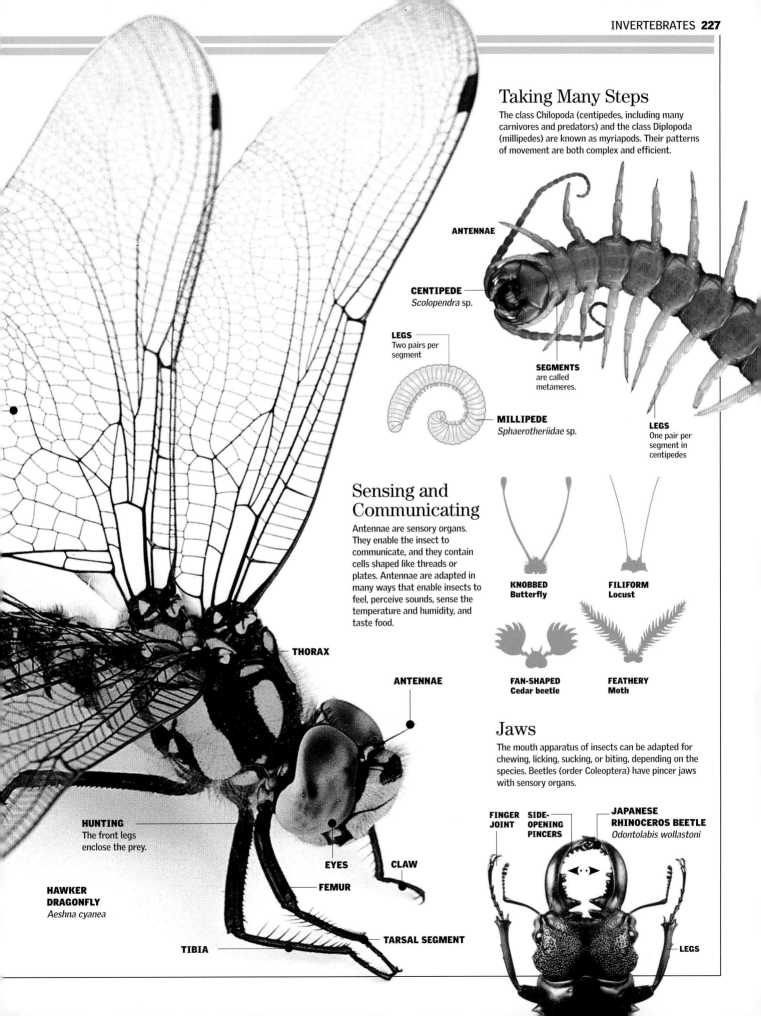

Taking Many Steps

The class Chilopoda (centipedes, including many carnivores and predators) and the class Diplopoda (millipedes) are known as myriapods. Their patterns of movement are both complex and efficient.

ANTENNAE

CENTIPEDE
Scolopendra sp.

LEGS
Two pairs per segment

SEGMENTS
are called metameres.

MILLIPEDE
Sphaerotheriidae sp.

LEGS
One pair per segment in centipedes

Sensing and Communicating

Antennae are sensory organs. They enable the insect to communicate, and they contain cells shaped like threads or plates. Antennae are adapted in many ways that enable insects to feel, perceive sounds, sense the temperature and humidity, and taste food.

KNOBBED
Butterfly

FILIFORM
Locust

FAN-SHAPED
Cedar beetle

FEATHERY
Moth

Jaws

The mouth apparatus of insects can be adapted for chewing, licking, sucking, or biting, depending on the species. Beetles (order Coleoptera) have pincer jaws with sensory organs.

FINGER JOINT **SIDE-OPENING PINCERS**

JAPANESE RHINOCEROS BEETLE
Odontolabis wollastoni

LEGS

THORAX

ANTENNAE

HUNTING
The front legs enclose the prey.

EYES

CLAW

FEMUR

HAWKER DRAGONFLY
Aeshna cyanea

TARSAL SEGMENT

TIBIA

The Better to See You With

Just as people without color vision have a hard time understanding what color is, it is impossible for humans to imagine what it is like to see through the compound eye of an insect. These eyes are made of thousands of tiny rods called ommatidia, each one a small eye connected directly to the brain. Scientists theorize that the insect's brain composes the images received from each ommatidium, enabling it to perceive movement in any possible direction—in some species, even from behind.

FLY
Drosophila sp.

Field of Vision

A fly's ommatidia are arranged in circles, and each one covers a portion of the field of vision. Such systems may not yield a high-resolution image, but they are highly sensitive to movement. The slightest motion causes a transfer of sensitivity from one ommatidium to another. This is what makes it so hard to catch a fly.

ANTENNA

360 degree
FIELD OF VISION OF A FLY

MOVING OBJECT

180 DEGREE FIELD OF VISION

BINOCULAR FIELD

SECTIONS OF FIELD OF VISION

35 INCHES (90 CM)

PERCEIVED PATH

HUMAN FIELD OF VISION

MOUTH
The mouth has apparatus for licking and sucking.

A Bee's Eye View

Compared with human vision, a bee's vision is somewhat nearsighted. Even the images of nearby objects are blurry. Its compound eyes have some 6,900 ommatidia.

DISTORTED MIDLINE

HUMANS
With binocular vision, a flat and undistorted image

BEES
In a larger field, the same image is narrower.

HEADED FOR NECTAR

Sensitivity to ultraviolet light, invisible to the human eye, enables worker bees to find the nectar inside the flowers.

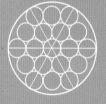

AREA WITH NECTAR

One Eye, or Thousands

Each ommatidium is responsible for a small portion of the visual field. Depending on the type of light they receive, the pigmented cells around each rhabdom can vary their diameter, regulating the overall sensitivity of the compound eye.

COMPOUND EYE

ANTENNA

OMMATIDIUM

THE HOUSEFLY HAS

4,000

OMMATIDIA.

RHABDOM
Connects each lens with its nerve

RETINAL CELL

LENS
Cone-shaped, to direct light to the rhabdom

PIGMENT CELL

CORNEA
Hexagonal in shape to fit into the rest of the compound eye

FLY VISION

EYELET

RETINA

OMMATIDIA

ANTENNA

TYPES OF EYES

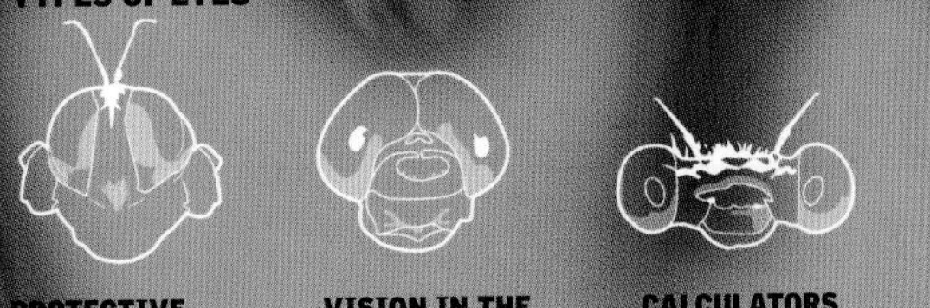

PROTECTIVE
The eyelets of the tachinid fly cover its eyes.

VISION IN THE ROUND
Certain dragonflies have a completely spherical field of vision.

CALCULATORS
This common blue damselfly uses its eyes to calculate distances.

Types of Mouths

Far from being a mere opening, the mouth is usually one of the most complex parts of an insect's body. The simple oral appendages of the most primitive forms were gradually modified so that this zoological group has been able to expand its diet. Thus, a hunter's mouth is totally different from that of a sucking insect or a leafeater, such as the locust.

ANTENNA

LOCUST
Family Acrididae
Since ancient times, locusts have been feared as a great plague on crops.

1 day
THE TIME IT TAKES A LOCUST TO EAT ITS OWN WEIGHT IN FOOD

Made to Order

The oral appendages of primitive insects were modified considerably, and they took different forms, according to the species. The first pair of upper jaws is for holding and sucking the food into the mouth. The second pair of upper jaws fuses at the midline during its development to form the lip, a structure with different functions, depending on the diet. The lower jaws and the first pair of upper jaws are at the side of the mouth, and an upper lip, the labrum, protects the front of the mouth. These parts form the basic biting-chewing apparatus. In more advanced forms, its modifications give rise to structures for sucking and licking or for biting and sucking.

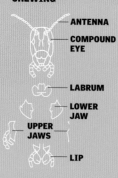

BITING AND CHEWING
- ANTENNA
- COMPOUND EYE
- LABRUM
- LOWER JAW
- UPPER JAWS
- LIP

LOCUST
Strong lower jaws and dexterous upper jaws

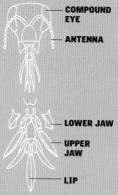

PIERCING AND CHEWING
- COMPOUND EYE
- ANTENNA
- LOWER JAW
- UPPER JAW
- LIP

BEE
Lip for nectar; lower jaws to chew pollen and mold wax

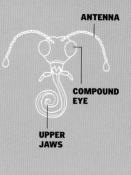

SUCKING
- ANTENNA
- COMPOUND EYE
- UPPER JAWS

BUTTERFLY
With a small labrum and no lower jaw. The upper jaws form a suction tube.

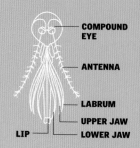

PUNCTURING AND SUCKING
- COMPOUND EYE
- ANTENNA
- LABRUM
- UPPER JAW
- LIP
- LOWER JAW

MOSQUITO (FEMALE)
The lip and upper jaws form a tube; the lower jaws are for puncturing the skin, and the labrum forms a sheath.

SEVEN-SPOTTED LADYBUG
Coccinella septempunctata feeds on aphids, plant lice, and sand flies.

Leaf Eaters

Insects, such as locusts and some beetles, as well as caterpillars (and the larvae of many other species), need a mouth structure capable of cutting leaves into small pieces and then putting them into the mouth. For this purpose, their large lower jaws have a series of serrated teeth, whereas the upper jaws and the lip have palps for manipulating and grasping the leaf pieces.

CARNIVORES
use their jaws as pincers to grasp their prey.

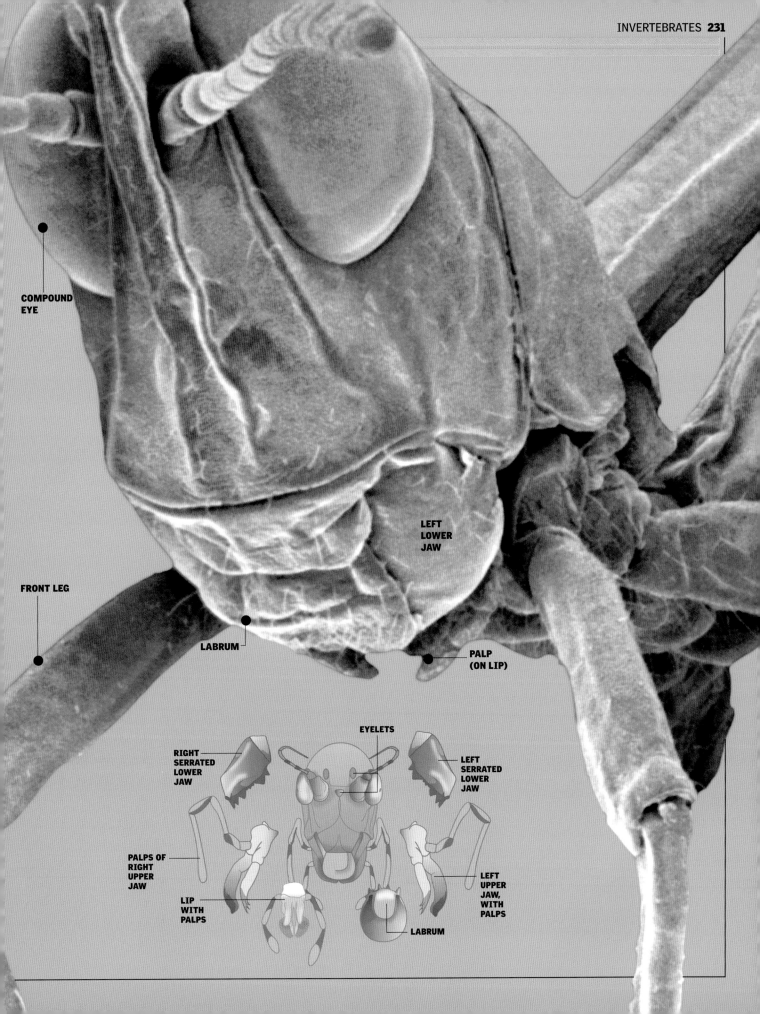

COMPOUND
EYE

FRONT LEG

LABRUM

LEFT
LOWER
JAW

PALP
(ON LIP)

RIGHT
SERRATED
LOWER
JAW

EYELETS

LEFT
SERRATED
LOWER
JAW

PALPS OF
RIGHT
UPPER
JAW

LIP
WITH
PALPS

LEFT
UPPER
JAW,
WITH
PALPS

LABRUM

Great Walkers

Etymologically speaking, myriapod means "many feet." The term refers to two very different classes of invertebrates: class Chilopoda and class Diplopoda, better known as centipedes and millipedes. All are animals divided into segments. Centipedes, most of which are carnivores, have a pair of legs on each segment, and millipedes have two pairs of legs per segment. These invertebrates (which are not insects) have so many legs that, to walk, they must use a highly sophisticated timing mechanism that seems to follow mathematical principles.

Applied Math

To walk, land-dwelling arthropods arch their bodies and move their six legs in coordination so that when one leg moves forward in a power stroke, the ones in front, behind, and opposite it are in a recovery phase, remaining on the ground. Myriapods have a similar mechanism, although it is much more complex because of their large number of legs. The legs are jointed, but they do not function independently of each other for the arthropod to move forward. The segmented body moves side to side in a regular wave pattern, and the legs are functionally adapted to this body movement.

SINUOUS PATH

Not only does an arthropod's body move in wave patterns; when the legs on one side of its body are closest together, those on the other side are farthest apart. This alternating pattern is repeated all along its body.

A Thousand Legs

Millipede, or diplopod, is the name of land-dwelling invertebrate species with multi-segmented bodies that have two pairs of legs on each segment. They live in damp places and feed on decomposing material. All of them have a pair of simple eyes and a pair of antennae, lower jaws, and upper jaws. The largest do not exceed 4 inches (10 cm) in length.

1 Like Claws

The back legs of this centipede, almost perpendicular to the rest, are used to trap and hold prey while it injects its catch with venom by means of powerful jaws.

2 Deadly Weapon

Holding the prey with the back legs, the centipede doubles itself over, the tips of the maxillipeds (forceps) fasten like claws, and the venom is discharged through the opening at the tip.

SIZE OR QUANTITY

There are species of centipedes that add segments throughout their lives; others are born with a fixed number of segments that grow in size.

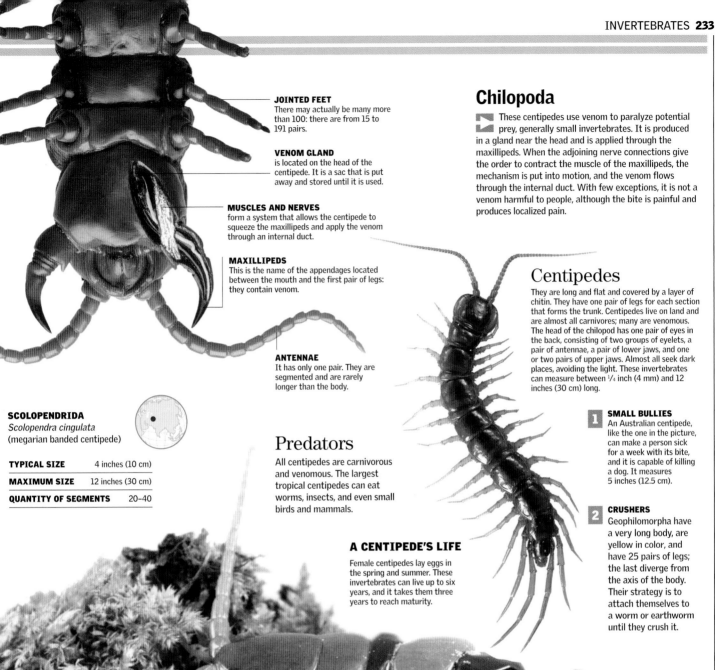

JOINTED FEET
There may actually be many more than 100: there are from 15 to 191 pairs.

VENOM GLAND
is located on the head of the centipede. It is a sac that is put away and stored until it is used.

MUSCLES AND NERVES
form a system that allows the centipede to squeeze the maxillipeds and apply the venom through an internal duct.

MAXILLIPEDS
This is the name of the appendages located between the mouth and the first pair of legs: they contain venom.

ANTENNAE
It has only one pair. They are segmented and are rarely longer than the body.

SCOLOPENDRIDA
Scolopendra cingulata
(megarian banded centipede)

TYPICAL SIZE	4 inches (10 cm)
MAXIMUM SIZE	12 inches (30 cm)
QUANTITY OF SEGMENTS	20–40

Chilopoda

These centipedes use venom to paralyze potential prey, generally small invertebrates. It is produced in a gland near the head and is applied through the maxillipeds. When the adjoining nerve connections give the order to contract the muscle of the maxillipeds, the mechanism is put into motion, and the venom flows through the internal duct. With few exceptions, it is not a venom harmful to people, although the bite is painful and produces localized pain.

Centipedes

They are long and flat and covered by a layer of chitin. They have one pair of legs for each section that forms the trunk. Centipedes live on land and are almost all carnivores; many are venomous. The head of the chilopod has one pair of eyes in the back, consisting of two groups of eyelets, a pair of antennae, a pair of lower jaws, and one or two pairs of upper jaws. Almost all seek dark places, avoiding the light. These invertebrates can measure between 1/4 inch (4 mm) and 12 inches (30 cm) long.

1 SMALL BULLIES
An Australian centipede, like the one in the picture, can make a person sick for a week with its bite, and it is capable of killing a dog. It measures 5 inches (12.5 cm).

2 CRUSHERS
Geophilomorpha have a very long body, are yellow in color, and have 25 pairs of legs; the last diverge from the axis of the body. Their strategy is to attach themselves to a worm or earthworm until they crush it.

Predators

All centipedes are carnivorous and venomous. The largest tropical centipedes can eat worms, insects, and even small birds and mammals.

A CENTIPEDE'S LIFE
Female centipedes lay eggs in the spring and summer. These invertebrates can live up to six years, and it takes them three years to reach maturity.

High-Quality Jumpers

Fleas are well known for their extraordinary jumps. When they are adults, these small, wingless insects take advantage of their jumping ability to hunt for their food, the blood of birds and mammals. They are ectoparasites of dogs, cats, and chickens, which keeps them present in our daily lives. They invariably bite their hosts and suck the blood that circulates through their skin.

FLEAS CAN SURVIVE

3 months

WITHOUT EATING.

Superprotein

The capacity to jump is related to the presence of resilin, a protein of great elasticity similar to rubber. Flea resilin has the function of building tension in the jumping legs. The release of accumulated energy generates the jump. On occasion, the jump is useless, and the flea does not manage to place itself on the host. Far from being a failure, the fall adds to the tension of the resilin, which makes the rebound a longer jump.

FLEAS IN THE HOME
Fleas are very common on dogs and cats. Fleabites generate serious discomfort for domestic animals, because scratching irritates and injures their skin.

② Action

The flea accumulates energy by tensing the muscles of the thorax and legs. When the accumulated elastic energy reaches a certain level, the flea releases its legs. As a result, the legs generate a sudden movement that causes the flea to jump.

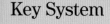

JUMPING LEG
The legs are furnished with extra upper segments. These supplements allow it to jump with speed.

Key System

1 The muscles in the coxa contract, generating enormous tension. The resistance to the tension is supported by the exoskeleton.

2 Once the jump is started, within thousandths of a second the direction, intensity, and orientation of the jump are all established by the torque that the muscles and leg segments create for the flea to complete its jump.

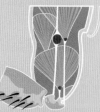

① Preparation

Within tenths of a second, the flea prepares itself to jump. It compresses the resilin and at the same time contracts its back legs. The back legs have a system of pads that retain the tension and accumulate energy.

Order of Jumpers

Fleas are in the class Insecta, order Siphonoptera, which includes wingless insects that are external parasites and lack wings. Their mouth apparatus is for piercing and sucking, and their life cycle is one of complete metamorphosis. Their 16 families include the genera of fleas that infest cats and dogs (*Ctenocephalides canis* and *C. felis*), as well as those that infest hens (*Ceratophyllus gallinae*).

DOG FLEAS
Ctenocephalides canis
This species is responsible for 90 percent of flea infestations in dogs.

HUMAN FLEA
Pulex irritans
usually feeds on human blood. Unlike other fleas, they do not remain on the host.

200 times
THE DISTANCE A FLEA CAN JUMP IN TERMS OF BODY LENGTH

③ In Flight

A flea can leap 24 inches (60 cm) in one bound. Its body is protected by armorlike overlapping plates that make up its exoskeleton. During a series of jumps, fleas can fall on their backs or heads without being injured.

Life Cycle

A complete cycle, from egg to adult, can take from two to eight months. The length of the cycle varies by species and by temperature, humidity in the environment, and the availability of food. In general, after feeding on blood, the female lays 20 eggs per day and up to 600 eggs throughout her life. The eggs are laid on the host (dogs, cats, rabbits, mice, rats, opossums, humans, etc.).

LARVA

EGGS

TOTAL CHANGE
Fleas are holometabolous; that is, their typical lifestyle includes a complete metamorphosis.

PUPA

ADULT FLEA

Edible Blood

As parasites of warm-blooded animals, fleas are classified as hematophagous (blood-eating) insects. Adults suck the blood of their hosts, which contains nutrients that they use for their own nutrition. Females use these nutrients to produce their eggs. The dried blood ejected in the adults' feces is also useful as food for various types of larvae.

1 The front legs are important for feeding. They hold the insect in place as it prepares to bite.

2 On injecting their stylet, fleas expel a substance that irritates the host but helps the fleas by keeping the blood from clotting while they are sucking it.

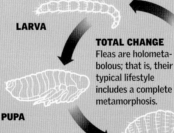

FLEA VS. MAN

A flea jumps a distance equivalent to 200 times the length of its body. To equal this feat, a man would have to jump over a 130-story building.

The Art of Flying

One of the most basic adaptations of insects has been their ability to fly. Most have two pairs of wings. Beetles (order Coleoptera) use one pair to fly and one pair for protection. For example, the rounded body of a ladybug is nothing more than the covering for a very sophisticated flight system. It makes these small beetles, which are harmless to humans, great hunters in the insect world.

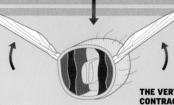

THE VERTICAL MUSCLE CONTRACTS AND THE WINGS MOVE UPWARD.

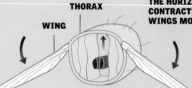

THORAX

WING

THE HORIZONTAL MUSCLE CONTRACTS AND THE WINGS MOVE DOWNWARD.

"Ladies" of Land and Air

 Some 4,500 species of these beetles live throughout the world. Almost all are brightly colored, with black spots on a red, yellow, or orange background. These colors warn away predators, who usually associate bright colors with poison. In fact, some ladybugs are actually poisonous for small predators, such as lizards and small birds. Ladybugs pose the greatest danger to agricultural pests, such as plant lice and gadflies, so they are often used as a natural biological pest control.

3

Flight

With the elytra open and spread like airplane wings, the second pair of wings is free to move. The muscles at their base control the direction of flight.

2

Takeoff

Although the colorful elytra are not used in flying, the insect needs to lift them in order to unfold its wings, which are seen only during flight.

Raised elytra

FRONT VIEW OF ELYTRA

40–80 inches per second
(1–2 m/s)
AVERAGE SPEED OF FLIGHT

SEVEN-SPOTTED LADYBUG
Coccinella septempunctata
Thanks to their help in destroying pests, during the Middle Ages these beetles were considered instruments of divine intervention from the Virgin Mary.

1

Preparation

The elytra can separate from the rest of the body. They protect the thorax, and also the wings when folded inside.

WINGS PREPARED FOR FLIGHT

The insect is between 1/25 and 1/2 inch (0.1–1 cm) long.

ELYTRON
Name of the modified front wing of beetles

RAISED ELYTRON

VISIBLE WING

BACK VIEW

APOSEMATISM
The opposite of mimetism: these insects use their bright colors to scare away danger.

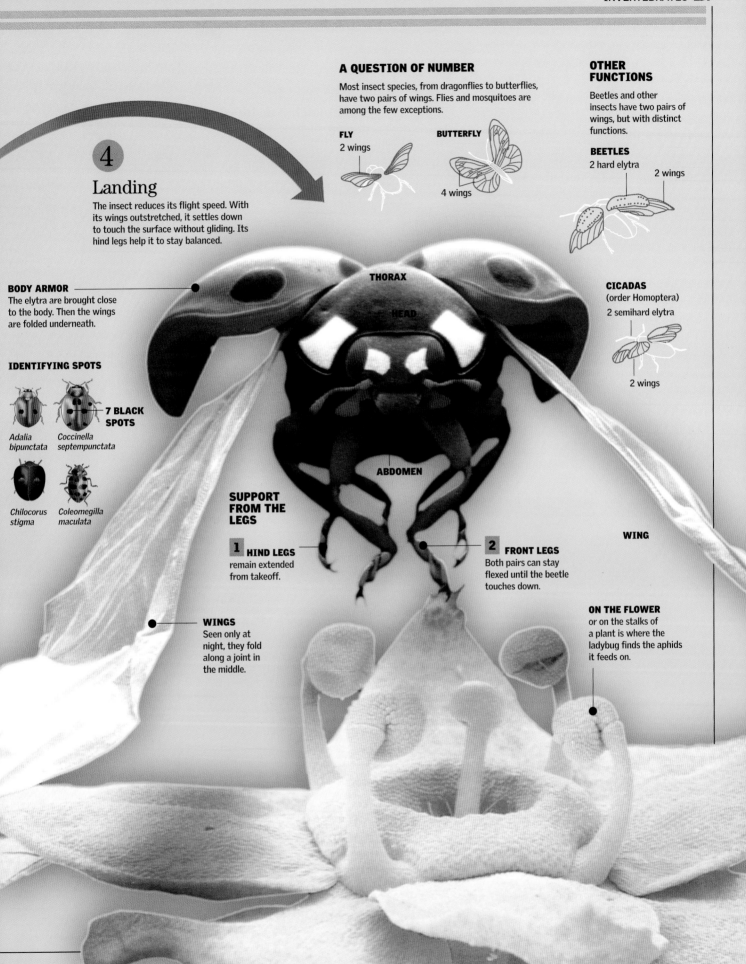

A QUESTION OF NUMBER

Most insect species, from dragonflies to butterflies, have two pairs of wings. Flies and mosquitoes are among the few exceptions.

FLY
2 wings

BUTTERFLY
4 wings

OTHER FUNCTIONS

Beetles and other insects have two pairs of wings, but with distinct functions.

BEETLES
2 hard elytra
2 wings

CICADAS
(order Homoptera)
2 semihard elytra
2 wings

4

Landing

The insect reduces its flight speed. With its wings outstretched, it settles down to touch the surface without gliding. Its hind legs help it to stay balanced.

THORAX

HEAD

ABDOMEN

BODY ARMOR
The elytra are brought close to the body. Then the wings are folded underneath.

IDENTIFYING SPOTS

Adalia bipunctata

Coccinella septempunctata

7 BLACK SPOTS

Chilocorus stigma

Coleomegilla maculata

SUPPORT FROM THE LEGS

1 **HIND LEGS**
remain extended from takeoff.

2 **FRONT LEGS**
Both pairs can stay flexed until the beetle touches down.

WING

WINGS
Seen only at night, they fold along a joint in the middle.

ON THE FLOWER
or on the stalks of a plant is where the ladybug finds the aphids it feeds on.

Metamorphosis

Metamorphosis is the change in shape that insects undergo as they grow. There are two types of transformations: complete, like that of monarch butterflies, and incomplete, like that of dragonflies or grasshoppers. Insects with complete metamorphosis pass through an immobile state (called the pupal, or chrysalid, phase) in which their body is transformed by hormones within a cocoon.

1 In the Beginning, the Egg

The adult female lays eggs among the leaves, where they will be protected. Monarch butterfly eggs have colors ranging from grayish white to cream, and they are shaped like barrels, 1/10 inch (2 mm) in diameter. The larvae grow inside the egg until they hatch; after hatching, they eat the shell.

MATING AND EGG LAYING

When monarch butterflies mate, they stay joined all afternoon and evening, until the next morning, for a total of 16 hours. After their first mating, the females lay eggs.

7 days
AMOUNT OF TIME THE LARVA LIVES INSIDE THE EGG

FIVE CHANGES

When it hatches, the insect is shaped like a worm. This caterpillar will molt its exoskeleton five times as it grows in size; however, its internal structure will not change. Each new exoskeleton is larger than the one before.

HATCHING FROM THE EGG
The exoskeleton hardens. As the insect grows, the exoskeleton becomes too small. Eventually, it splits and falls off.

SECOND SHEDDING

THIRD SHEDDING

FOURTH SHEDDING

CHANGE TO PUPAL PHASE

MARIPOSA MONARCA
Danaus plexippus

Simple Metamorphosis

Also called incomplete metamorphosis, because, unlike complete metamorphosis, it does not include a pupal phase. The wings and legs develop gradually, so that the insect does not need to spend a certain amount of time immobile. Locusts, cockroaches, termites, and dragonflies have this type of metamorphosis. From an evolutionary standpoint, it corresponds to ancient or primitive insects. One of its characteristics is the nymph stage of young insects. The nymph gradually changes in shape as it grows. When it sheds its exoskeleton, the adult emerges.

EMPEROR DRAGONFLY
Anax imperator

1 EGG

2 NYMPH

3 IMAGO (ADULT)

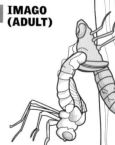

Larva or Caterpillari

makes its entry into the world by eating its shell. From then on, eating and growing will be its main activities. Every time it sheds its skin, the old exoskeleton is broken. The insect forms a new, soft exoskeleton, which is gradually expanded by blood pressure. The exoskeleton then undergoes a chemical reaction that hardens it.

3 weeks
IS THE AMOUNT OF TIME THE INSECT LIVES IN THE LARVAL STAGE.

A SIMPLE ASSIGNMENT

In the caterpillar phase, the insect focuses solely on eating leaves. In this way it accumulates the necessary energy for the physiological processes of metamorphosis. For digesting the leaves, the caterpillar has a very simple digestive track.

PREPARATION FOR THE PUPAL PHASE

Before passing to the next stage, the larva stops eating and eliminates any food left in its digestive tract. The juvenile hormone, which keeps the transformation of the body in check, starts to become inhibited.

CREMASTER
The caterpillar secretes a fibrous cushion that sticks to the stalk of a plant. It hangs from the cushion with hooks on the end of its abdomen.

HANGING AND IMMOBILE
To leave the larval stage behind and become a pupa, the caterpillar quietly awaits the transformation.

EXOSKELETON
Crossed with yellow, black, and white stripes, it is soft after every shedding and later hardens. The insect always emerges head first.

INSIDE THE LARVA
The insect's heart, nervous system, and breathing system are almost completely developed during the larval stage, and they change very little afterward. The reproductive system is formed later.

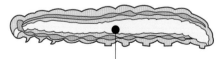

INTESTINE

FAREWELL TO THE OLD BODY
The larva's last exoskeleton begins to fall off and is replaced by a greenish tissue that will form the cocoon, or chrysalis.

③ Pupa (Chrysalis)

After getting rid of its larval exoskeleton, the insect hangs immobile from a branch, protected by a cocoon. Inside, it will develop its distinctive butterfly form. Although it does not eat during this period, it is intensely active biologically and undergoes considerable change. Histolysis, a process in which the larva's structures are transformed into the material that the insect will use to develop adult structures, takes place at this time.

HISTOGENESIS

New tissues are generated from hemolymph (the equivalent of blood), the Malpighian tubules (the energy-producing organ in insects), and histolyzed tissue, including the larva's muscles. The monarch butterfly pupa is called a chrysalis because of the color and structure of the capsule that protects it. It is oval-shaped with gold and black spots.

15 days

LENGTH OF THE PUPA, OR CHRYSALIS, PHASE

BUTTERFLY SHAPE

The adult butterfly's wings and legs develop from the cuticle, or skin tissue, which is composed mostly of chitin. Other organs are preserved or rebuilt from regenerative cells.

HORMONES IN FULL SWING

Metamorphosis is governed by three hormones. One is the cerebral (brain) hormone, which stimulates the prothoracic gland. This gland produces the molting hormone ecdysone, which causes the loss of the old skin. The third hormone is the juvenile hormone, which slows down the transformation to the adult stage.

WITHIN SIGHT

As the time draws near for the adult to emerge, the chrysalis becomes thinner, changes color, and becomes transparent. The transformed insect can be seen inside.

CAMOUFLAGE
The chrysalid capsule has shapes, textures, and colors that help keep it from being noticed to protect it from predators. The capsules typically resemble leaves or bird droppings.

INTERNAL ORGANS
Inside the chrysalis, the insect's body is changing into that of an adult. The intestine rolls into a spiral shape to assimilate liquid food, and the reproductive organs are developing for the adult stage.

INTESTINE

ANATOMY OF A BUTTERFLY

The body of a butterfly is divided into head, thorax, and abdomen. An adult butterfly's head has four important structures: eyes, antennae, palps, and proboscis. The compound eyes of a butterfly are made up of thousands of ommatidia, each of which perceives light. There are two antennae and the two palps, and they are covered by scales that detect molecules in the air and give the butterfly a sense of smell. The proboscis is a modified tongue through which the butterfly draws nectar and water to feed itself, and it is rolled up when not in use. The thorax is made up of three segments, each of which has a pair of legs attached to it. The second and third segments are also each joined to a pair of wings. Each leg is formed by six segments. When a butterfly lands on a plant, it uses the end segments, called tarsi, to grip the surface of a leaf or flower.

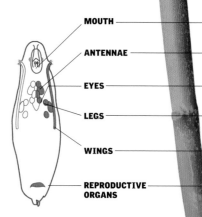

MOUTH

ANTENNAE

EYES

LEGS

WINGS

REPRODUCTIVE ORGANS

④ Adult

After reaching its final shape, the butterfly will not grow anymore. When the butterfly emerges from its cocoon, its wings are still wrinkled and damp. It will need to hang upside down to stretch them out to dry, so that they will be useful for flying. This will take several hours of waiting and struggling. From then on, the butterfly will feed on nectar.

BREAKING FREE
To permit the butterfly to emerge, the mature capsule splits along its length. The insect gradually stretches its new body and activates the circulation of the hemolymph.

FLY AWAY, BUTTERFLY
The life span of this insect in its adult phase will depend on its luck, its migrations, and the attacks of predators . . .

from 5 to 7 weeks

YOUNG ADULT
Once free of its covering, the adult is usually pale in color, and its wings are soft and folded. After about 40 minutes the wings expand, harden, and take on their full color.

ELIMINATING WASTES
While emerging, the butterfly secretes a fluid containing the waste produced during the chrysalis phase. This fluid, called meconium, is considered rather foul-smelling.

JOINED, BUT NOT FOREVER
After leaving the chrysalis, monarch butterflies from the same litter stay together for a period of three to eight days before they go their separate ways.

UPS AND DOWNS OF LIFE AS AN ADULT
Mating, reproducing, and laying tiny eggs to give rise to new generations will be the main activities of the adult insect. Each female lays an average of 100–300 eggs during her life.

Order and Progress

A nts are one of the insects with the highest social organization. In the anthill, each inhabitant has a job to do. The head of the family is the queen, the only one that reproduces. All the rest of the ants are her offspring. During mating, queens and drones (males) from various colonies mate on the wing. The queens need to mate several times, because the sperm they receive will have to last their lifetime.

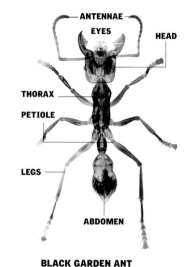

ANTENNAE
EYES
HEAD
THORAX
PETIOLE
LEGS
ABDOMEN

BLACK GARDEN ANT
Lasius niger

MAIN ENTRANCE

The Anthill

After mating, the queen loses her wings and chooses a place to lay eggs. At first she lives on reserves derived from the muscle mass of her wings and some of the first eggs she has laid. She takes charge of raising the first generation of worker ants, which will then take care of finding food while the queen focuses exclusively on laying eggs.

COMMUNICATION

An ant communicates with its antennae through chemical means, by capturing particles of certain substances (pheromones) that enable it to recognize another ant from the same colony. Ants do not have a well-developed sense for perceiving sound.

FOOD STORAGE
Honeypot ants coordinate the food supply.

METAMORPHOSIS

In the egg stage, the future ant remains near the queen but leaves her during the larval stage. Other ants then take care of the larva, and it will become a nymph and form a cocoon to cover itself.

UNUSED TUNNEL

THERE ARE ABOUT
10,000
ANT SPECIES.

EGGS   LARVAE   NYMPHS   COCOON

3 NYMPHS
are fed and taken care of in another area.

2 LARVAE
are carried to another chamber to grow.

1 EGGS
are laid by the queen in the lowest area.

4 COCOONS
The new ants hatch ready to work.

QUEEN ANT

 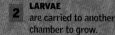

YOUNG ANTS

The Castes

Each ant plays a role in the nest and is assigned its role at birth. Drone, soldier, worker, and replete worker (which stores food reserves) are the castes that distinguish what chores each ant will have.

Four wings

QUEEN
The largest ant. She lays the eggs that will become workers, drones, and new queens.

Two wings

DRONE
His only function is mating; afterward he dies.

WORKER
The worker ant may have the role of gathering food, cleaning, or protecting the anthill.

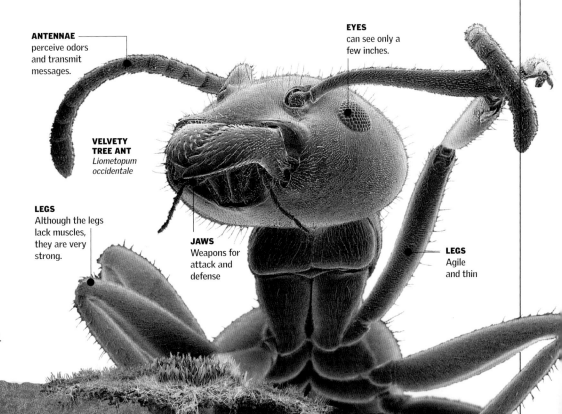

ANTENNAE
perceive odors and transmit messages.

EYES
can see only a few inches.

VELVETY TREE ANT
Liometopum occidentale

LEGS
Although the legs lack muscles, they are very strong.

JAWS
Weapons for attack and defense

LEGS
Agile and thin

Feeding

Ants cannot eat solid food. The plants and animals they eat are mixed with saliva to form a paste, which is used to feed the whole colony.

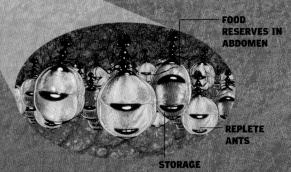

FOOD RESERVES IN ABDOMEN

REPLETE ANTS

STORAGE

Defense

The most widely used defense is biting and spraying streams of formic acid. Soldier ants have the job of scaring away the enemy because they have larger heads than worker ants.

JAW

The jaw is the ant's main weapon of defense, with a bite that can scare away or harm a rival. The jaw is also used for hunting and feeding.

AMERICAN FARMER ANT

CLAMPING JAW

VENOM

may contain formic acid and can kill or paralyze the prey. It comes from special glands in the lower abdomen.

INTERCHANGE OF FOOD

Having two stomachs, an ant can share food. The transfer begins when the receiving ant uses its front legs to touch the lip of the donor ant.

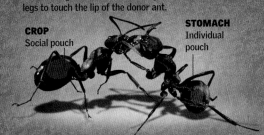

CROP
Social pouch

STOMACH
Individual pouch

SOUTHERN WOOD ANT
Formica rufa

POISONOUS STINGER

ABDOMEN

STREAM OF POISON

TRAP-JAW ANT
Odontomachus bauri

POISON SAC

Goal: Survival

Evolution has molded some striking traits into living beings. In particular, some insects, disguised as branches or leaves, can escape notice so it can hunt or hide from predators. To avoid being attacked, other insects develop colors and shapes that deceive other animals and keep them from attacking. Hiding and showing off are two opposite strategies that have been favoring the survival of the fittest for millions of years.

BRIMSTONE BUTTERFLY
Gonepteryx sp.
The profile of the wings resembles the shape of cut leaves.

PEACOCK BUTTERFLY
Inachis io
The flashy, aposematic (warning) coloration keeps predators away by warning of the danger the insect poses.

WINGS
These wings look like leaves, with a similar color, shape, and structure.

FALSE EYE
The scales are pigmented to look like eyes.

Masters of Simulation

Camouflage, or crypsis, is a phenomenon in which animals use amazing disguises as advantageous adaptations. Camouflage is used both by hunters and by potential prey. Insects' bodies may be disguised as various substrates and parts of trees, such as bark, leaves, and branches. These masking techniques are a convenient way for the insect to fade into the background.

Disguise

These insects use survival strategies designed to keep predators from seeing them. This disguise is their only means of defense.

DOUBLE PROTECTION
Caligo sp.
Owl butterflies combine Batesian and Mullerian mimicry. Predators can confuse the owl butterfly with leaves, but if a predator succeeds in finding it, the butterfly folds its wings to look like the shape and eyes of an owl. The predator, confused, backs off from attacking.

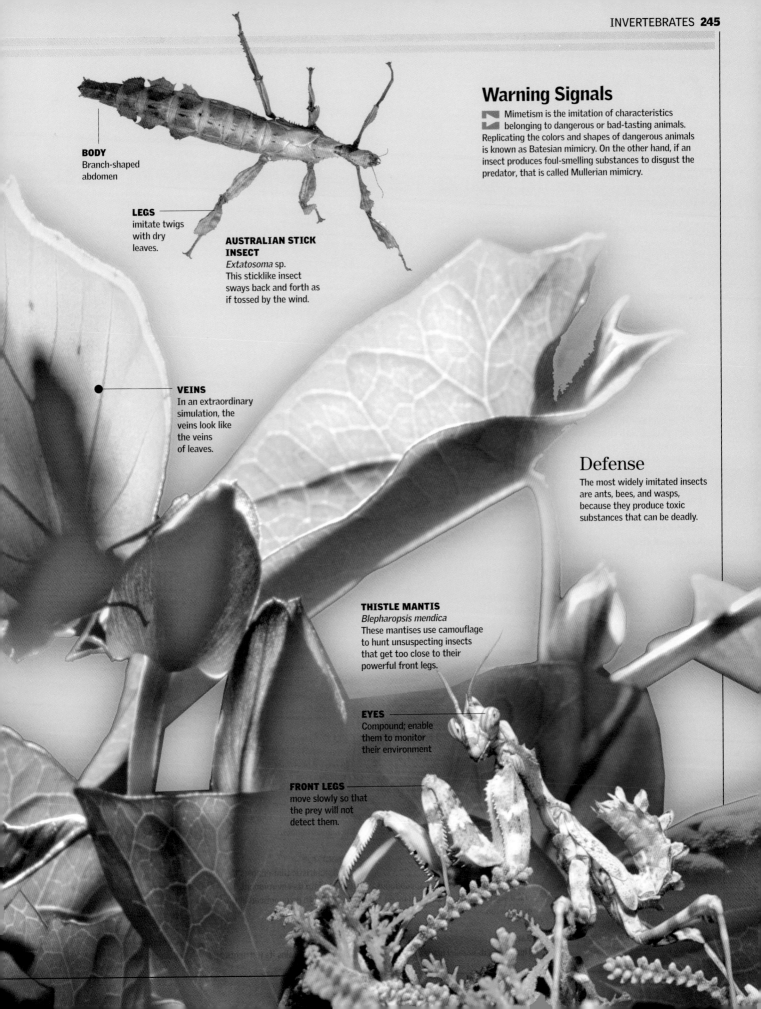

BODY
Branch-shaped
abdomen

LEGS
imitate twigs
with dry
leaves.

**AUSTRALIAN STICK
INSECT**
Extatosoma sp.
This sticklike insect
sways back and forth as
if tossed by the wind.

Warning Signals

Mimetism is the imitation of characteristics
belonging to dangerous or bad-tasting animals.
Replicating the colors and shapes of dangerous animals
is known as Batesian mimicry. On the other hand, if an
insect produces foul-smelling substances to disgust the
predator, that is called Mullerian mimicry.

VEINS
In an extraordinary
simulation, the
veins look like
the veins
of leaves.

Defense

The most widely imitated insects
are ants, bees, and wasps,
because they produce toxic
substances that can be deadly.

THISTLE MANTIS
Blepharopsis mendica
These mantises use camouflage
to hunt unsuspecting insects
that get too close to their
powerful front legs.

EYES
Compound; enable
them to monitor
their environment

FRONT LEGS
move slowly so that
the prey will not
detect them.

GLOSSARY

Abdomen
Posterior portion of the body of arthropods consisting of similarly formed segments, containing the reproductive organs and a part of the alimentary canal. In insects and arachnids, it is the posterior section of the body.

Abomasum
Last of the four chambers into which ruminants' stomachs are divided. It secretes strong acids and many digestive enzymes.

Abyssal Fish
Rare species that inhabit depths of 8,200 feet (2,500 m) and below, where no light reaches. They have peculiar shapes, with large heads and strong teeth for eating other fish, because no vegetation grows at those depths. They attract prey with lure organs consisting of photophores that shine in the darkness.

Actinopterygii (Ray-Finned Fish)
Class of fish distinguished mainly by having a skeleton with bony spines in the fins. They have a cartilaginous skull and only one pair of gill openings covered by an operculum.

Adaptation
A structural, physiological, or behavioral trait that allows an organism to live in its environment.

Aerodynamic
Having an appropriate shape to decrease resistance to the air.

Albumin
Protein found in abundance in blood plasma. It is the principal protein in the blood and is synthesized in the liver. It is also found in egg whites and in milk.

Alulae
Rigid feathers whose function is to decrease air turbulence during flight.

Ambulacral Groove
In echinoderms, any of the radial grooves through which the hydraulic system's tube feet protrude.

Amino Acid
Organic molecule from which proteins are produced.

Amphibians
The young live in the water, and the adults live on land. Many need to stay near water or in damp places to avoid drying out. This is because some species breathe mainly through their skin, which can absorb air only when damp. This group today includes frogs, toads, salamanders, and limbless caecilians.

Ampullae of Lorenzini
Organs in sharks for detecting signals emitted by potential prey.

Anaerobic
Breathing process that does not require oxygen.

Anal Fin
Unpaired fin located in the middle ventral part of the fish above the anus.

Ancestor
Progenitor, more or less remote, that passes down a set of characteristics to its descendants.

Anguilliformes
Fish with a long, slender body without appendages, including eels and morays.

Annelids
Animals with a long cylindrical body consisting of ring-formed segments.

Antennae
A pair of long sensory appendages on the head of many arthropods.

Aorta
Main artery in blood circulation systems. It sends blood to other tissues of the body.

Arachnid
An eight-legged arthropod.

Arthropod
An animal with articulated appendages and a segmented body, covered by an exoskeleton.

Barbs
Thin, straight, parallel blades, perpendicular to the shaft. They resemble the leaves of a palm tree.

Bathypelagic
Fish that live at ocean depths below the mesopelagic zone, where light cannot penetrate.

Bilateral Symmetry
Corporal form whereby the right and left halves of an organism are approximate mirror images of each other.

Biodiversity
Variety of species that live in a given natural or artificial environment.

Biogeographic Regions
Geographic regions that biologists analyze to determine the distribution of animals and other living organisms, according to the geographic conditions of a place. Migratory birds usually travel through different biogeographic regions between winter and summer.

Bioluminescence
Property of living beings that can produce light.

Biped
Animal that stands upright, walks, or runs using only the two hind limbs.

Bony Fish
Fish with bony skeletons and jaws. Their skeletons are relatively small but firm. They have flexible fins that allow precise control of their movements.

Bony Plates
Formations that grow from the skin and have a protective function for certain species.

Calcite
A form of the chemical compound calcium carbonate.

Camouflage
A characteristic that enables the animal to blend into its environment. It allows the animal to go unnoticed in the presence of predators.

Carnivore
Animal that obtains its nutrients and energy from a diet consisting mainly or exclusively of animal flesh.

Carnassial
A typical sharp premolar present in carnivorous animals that helps them cut and tear the flesh of prey more efficiently.

Carpus
Bone structure of the wrist, located between the bones of the forearm and the metacarpus. It is made up of two rows of bones.

Cartilaginous Fish
Fish with skeletons made of cartilage, such as the Elasmobranchii, a group that includes sharks and rays.

Carrion
The remains of dead animals used as food by some birds or other animals.

Caste
A social group that carries out specific tasks, characteristic of ants and bees, among other insects.

Caudal Fin
Unpaired fin at the lower end of the body, forming the tail fin in most fish.

Cephalopod
A class of exclusively marine mollusks with tentacles or legs attached to the head. These appendages have rows of suckers that are used for capturing prey and copulation.

Cephalothorax
The head and thorax combined in one single body segment.

Cerebellum
A section of the brain in vertebrates located above the brain stem and behind and below the cerebrum. It coordinates muscular activity and maintains balance.

Chelicera
First pair of appendages in crabs, sea spiders, and arachnids, usually in the form of pincers or fangs.

Chelonia
Collective term for land and sea turtles.

Chick
A baby bird that has just come out of the eggshell and that has not yet left the nest. Its diet and safety depend on its parents.

Chiridium
A muscular limb in tetrapods. It is a long bone whose anterior end articulates with the scapular belt. The posterior end articulates with two bones that connect to the joints of the digits.

Chitin
Tough, durable polysaccharide that contains nitrogen and is found in the exoskeleton of arthropods or other surface structures of many invertebrates, and also in the cell walls of fungi.

Chordate
Animal that belongs to the phylum Chordata; any animal having a spinal cord, whether throughout its development or only in certain stages. Animals that are not chordates are called invertebrates.

Class
One of the many divisions scientists use to classify animals.

Classification
The process of establishing, defining, and ordering taxa within a hierarchical series of groups.

Climate
Average temperature, humidity, and pressure that determine the atmospheric conditions of a region and that are related to other geographic characteristics of that region.

Cloaca
The open chamber into which the ducts of the urinary and reproductive systems empty.

Cochlea
A structure shaped like a coiled spiral tube, located in the inner ear of mammals.

Cocoon
A protective sheath usually made of silk. Many insects make cocoons to protect themselves during the pupa stage, until they become adults.

Cold-Blooded
Organism whose body temperature is mainly controlled by an external heat source because it has little capacity to generate its own heat.

Colony
A group of animals of the same species that live and work together to survive.

Community
The entire population of organisms that inhabit an environment in common and who interact with one another.

Complete Metamorphosis
Phenomenon where the adult form of an animal looks nothing like the immature form; examples are frogs and toads.

Compound Eye
In arthropods, a complex eye made of many separate units, each of which has light-sensitive cells and a lens that can form an image.

Cones
The photoreceptor cells in the retina of vertebrates. They are essential for distinguishing colors.

Connective Tissue
Tissue that joins, supports, and protects the other three types of tissues: epithelial, muscular, and nervous. It contains a network composed of many fibers surrounding the cells.

Convolution
Each of the slight elevations or folds that mark the surface of the cerebral cortex.

Cortex
The outer tissue of some organs, such as the brain and kidney.

Crest
Extended or raised feathers located on the upper part of a bird's head.

Crustacean
An animal of the arthropod group, with antennae and articulated appendages, that uses gills to breathe and has a body protected by a thick covering.

Crop
Membranous sac that communicates with a bird's esophagus, where food is softened.

Ctenoid
Type of scale in which the free edge has spines.

Cuticle
An organic, noncellular tough but flexible protective outer covering secreted by the epidermis.

Cycloid
Type of scale in which the free edge is rounded.

Dendrite
The branched elongation of a nerve cell by means of which it receives external stimuli.

Dermal Papilla
Structure from which a feather develops. It is composed of epidermal and dermal cells.

Dermis
The inner layer of the skin, located under the epidermis.

Dewlap
Fold of skin hanging below the chin and extending to the chest in some lizards and other tetrapods. It can be unfolded in territorial battles to intimidate or to display certain moods.

Digitigrade
Refers to animals that use only their digits to walk. One example is dogs.

Display
Behavior directed at attracting the attention of a partner. It can also be done to threaten or distract predators.

Distribution
Place where a species is located. It includes the area the species occupies in different seasons.

Diversity
Degree to which the total number of individual organisms in an ecosystem is distributed among different species. Minimum diversity is reached when all the organisms belong to one species. Maximum diversity is reached in stable natural environments with a maximum variation in the substrate and environmental conditions.

DNA
Deoxyribonucleic acid. Double-helix shaped molecule that contains encoded genetic information.

Dorsal Fin
Unpaired fin located on the back, which keeps the fish in a stable position.

Down Feather
A very thin and light feather, similar to silk, that birds have underneath their external plumage. Down feathers constitute the first plumage of chicks.

Duvernoy's Glands
System possessed by some snakes for injecting venom. They are a pair of modified salivary glands, one on each side of the head.

Echinoderms
Invertebrate marine animals. The bodies of the adults have a pentagonal symmetry. Underneath the skin they have a calcareous skeleton with spines and protuberances. They have an internal hydraulic system, connected with ambulacral feet, that makes locomotion possible.

Echolocation
The ability to orient and maneuver by emitting sounds and interpreting their echoes.

Ecosystem
A dynamic system formed by a group of interrelated living beings and their environment.

Egg
Fertilized ovule that develops into a new individual. It usually also refers to the entire structure that covers and protects the fertilized ovule.

Egg Tooth
Sharp calcium growth, in the shape of a tooth, that forms on the tip of a chick's bill during the embryonic phase. The chick uses the egg tooth to break its shell at birth.

Embryo
The first stage of development of a multicellular animal or plant.

Endemic
Native to a particular geographical region and restricted to it.

Endoderm
One of the three layers of the embryonic tissue of animals; it originates in the epithelium that covers certain internal structures.

Environment
The natural conditions, such as vegetation and land, that influences the development and behavior of animals.

Epidermis
The outer layer of the skin formed by epithelial tissue covering the bodies of animals.

Epipelagic
Relating to organisms that live in open water away from the ocean floor, from the surface to depths of approximately 660 feet (200 m).

Evolution
Gradual process of change that a species undergoes to adjust to the environment

Exoskeleton
The external covering supporting the body, commonly found in arthropods. It is like an articulated shell made of chitin; it serves as a support for muscles and the soft internal organs.

External Fertilization
Fertilization of eggs that takes place outside the female's body. The male releases sperm over the eggs after the female deposits them. The eggs are exposed to the outer environment.

Extinct
No longer existing. Many bird species are now extinct (for example, ictiornites).

Eyelet
Simple light receptor, common among invertebrates.

Family
A category in taxonomy that groups genus together; lower than order and higher than genus.

Feather
Each unit of the covering (plumage) of a bird. The feathers are composed of a hard substance called keratin. They have a long quill, to which two blades are joined. The blades—formed by a great number of barbs, uniformly distributed—give the feather its shape and color.

Fertilization
Union of the reproductive cells of a male and a female that will create a new individual.

Flying Fish
Exocoetids, or flying fish, are a family of 70 species of ocean fish in nine genera. They are found in all the oceans, especially in warm subtropical and tropical waters. Their most notable characteristic is their unusually large pectoral fins, which enable them to glide through the air for short distances.

Follicle
A small organ in the form of a sac located in the skin or mucous membranes.

Fossil
Remains of various types of ancient life-forms, both plants and animals, in a rocky substrate. They are found in the geological strata of the earth's surface.

Fry
Newly hatched fish whose shape resembles that of adults of the same species.

Gamete
The mature reproductive cell that combines with a gamete of the opposite sex to form a zygote that is usually diploid; male gametes are called spermatozoids and female gametes are called ovules.

Ganoid
Type of scale made of shiny, enamel-like material (ganoin) formed in successive layers over compact bone. The extinct fish Palaeospondylus had this type of scale. The only modern fish with ganoid scales are gar, bowfin, and reedfish.

Gastric Juice
Set of fluids produced by the stomach glands of birds and other animals.

Gastrovascular Cavity
A digestive cavity with an opening, characteristic of the phyla Cnidaria and Ctenophora. It has digestive and circulatory functions.

Gene
Unit of information in a chromosome; sequence of nucleotides in the DNA molecule that carries out a specific function.

Genus
A category in taxonomy that groups species together.

Gestation
The state of an embryo inside a woman or female mammal from conception until birth.

Gill Arch
Bone that anchors the gill filaments or spines.

Gills
Respiratory organs of aquatic animals.

Often an extension of fine tissues from the outer surface of the body or, in vertebrates, from part of the digestive tract.

Gizzard
Muscular stomach of a bird. It is very robust, especially in granivores, and it is used to grind and soften the food by means of mechanical pressure. The food arrives at the gizzard mixed with digestive juices.

Gland
Group of epithelial cells that produce secretions, organized inside a covering membrane to form an organ whose function is to synthesize and excrete molecules that the organ itself does not use.

Gonad
Glands that produce reproductive sex cells.

Habitat
The set of geophysical conditions in which an individual species or a community of animals or plants lives.

Hatching
Cracking of the eggshell so that the bird can come out.

Herbivore
Animal that feeds exclusively on plants.

Hermaphrodite
An organism that has both reproductive systems, male and female; hermaphrodites may or may not self-fertilize.

Heterocercal
Type of tail fin in which the spine curves upward, forming an upper lobe of larger size.

Hibernation
The physiological state that occurs in certain mammals as an adaptation to extreme winter conditions, exhibited as a drop in body temperature and a general decrease in metabolic function.

Homeostasis
The set of self-regulating phenomena that keeps the composition and properties of an organism's internal environment constant.

Homeothermy
Thermoregulation characteristic of animals that maintain a constant internal temperature, regardless of external conditions. Body temperature is usually higher than that of the immediate environment.

Hoof
Horny, or cornified, covering that completely envelops the distal extremity of horses' feet.

Hormone
An organic molecule, secreted in small amounts by one part of an organism, that regulates the function of other tissue or organs.

Host
An organism in which a parasite lives.

Hydrostatic Skeleton
A skeleton in which fluid is contained by muscular walls that transfer the force from one part of the body to another when subjected to pressure.

Incubation
The act of keeping the eggs warm so that the embryos inside can grow and hatch. Usually the chick's parents use their own bodies to warm the eggs, but some birds use sand or decomposing plants to cover them.

Instinct
Innate behavior that an animal develops and that is not learned. The offspring of ducks start to swim by instinct, for instance.

Internal Fertilization
Fertilization of cartilaginous fish, aided by the male's copulating organ. These organs, called claspers, developed from modifications of the pelvic fins.

Invertebrate
Animal without a spinal column. Some, such as worms, have soft bodies. Others, such as arthropods, are protected by a hard exoskeleton.

Iris
The membranous disk of the eye between the cornea and the lens that can take on different coloration. In its center is the pupil, which is dilated and contracted by the muscle fibers of the iris.

Jacobson's Organ
Organ on the upper part of the palate that takes in substances captured by a reptile's tongue and analyzes them to determine various characteristics of the object they come from. Also called the vomeronasal organ.

Keel
Ridge or fleshy border along the sides of the caudal peduncle.

Keratin
A protein rich in sulfur, it constitutes the chief element of the outermost layers of mammals' epidermises, including hair, horns, nails, and hooves. It is the source of their strength and hardness.

Kingdom
Taxonomic category that includes phyla or divisions. Until the appearance of the category of domain, the kingdom was the highest-level category in biological classification.

Lactation
The period in mammals' lives when they feed solely on maternal milk.

Larva
Animal in a developmental stage, after leaving the egg. It can feed itself but has not yet acquired the shape and structure of the adults of its species.

Lateral Line
Line along the sides of the fish's body consisting of a series of pores.

Lipids
Group of water-insoluble substances, including fats, oils, waxes, steroids, glycolipids, phospholipids, and carotenes.

Litter
All the offspring of a mammal born at one time.

Luminous Organs
Most fish in the ocean depths have bioluminescent organs that shine in the darkness and are used to attract prey or to communicate.

Lungfish
Fish that appeared in the Mesozoic era, 250 million years ago. Like amphibians, these species breathe with lungs and are considered living fossils. Only three species have survived to the present.

Mammals
Vertebrate animals whose females have mammary glands, which secrete substances that serve as food for their young.

Mammary Gland
One of a pair of external secretion organs characteristic of mammals.

It provides milk to the young during lactation.

Mandible
Appendage immediately below the antennae, used to trap, hold, bite, or chew food.

Mantle
In mollusks, the outer layer of the body wall or a soft extension of it. It usually secretes a shell.

Marsupial
Mammals whose females give birth to unviable infants, which are then incubated in the ventral pouch, where the mammary glands are located. They belong to the Metatheria infraclass.

Marsupium
The pouch, characteristic of female marsupials, that functions as an incubation chamber. It is formed by a fold of the skin and is attached to the outer ventral wall. The mammary glands are found there, and the offspring complete the gestation period there.

Medium
Element or substrate where organisms live.

Mesoderm
The middle layer of the three layers of embryonic tissue.

Mesopelagic
Relating to organisms that live in the ocean depths, where light is dim. The mesopelagic zone is intermediate between the upper or euphotic (well-lit) zone and the lower or aphotic (lightless) zone.

Metabolism
The sum of all the physical and chemical transformations that occur within a cell or organism.

Metacarpus
The set of elongated bones that make up the skeleton of the anterior limbs of certain animals and of the human hand. They are articulated to the bones of the carpus, or wrist, and the phalanges.

Metamorphosis
Abrupt transition from the larval form to the adult form.

Microorganism
Organism that can be seen only with a microscope.

Migration
Seasonal travel of animals from one region to another to reproduce or to seek food, better climate, or better living conditions in general.

Mimetism
Property of certain animals and plants to resemble living things or inanimate objects that live nearby, mostly by means of color.

Mimicry
Ability of certain organisms to modify their appearance to resemble elements of their habitat or other, better protected species, using camouflage to hide from their predators or prey.

Molars
Group of teeth that crush food within the mouth.

Mollusk
Invertebrates of the phylum Mollusca, with a soft body divided into a head, foot, and visceral mass. They have a fold called a mantle that envelops all or part of the body.

Molting
Removal of all or part of the outer covering of an organism; in arthropods, a periodic changing of the exoskeleton that enables them to grow in size.

Monogamous
Birds that mate with only one individual of the opposite sex. Many penguins have monogamous behavior.

Morphology
Study of the form of an object or structure. For instance, the morphology of the feet of birds is an area of study.

Mouth Incubation
Mode of gestation for certain fish species that incubate the eggs inside their mouth and spit them into a burrow to feed. When the eggs hatch, the parent protects the young inside its mouth.

Nectar
Sweet and sugary secretions found in flowers that attract birds and other animals. Hummingbirds feed on nectar.

Neuron
A differentiated cell of the nervous system capable of transmitting nerve impulses among other neurons. It is composed of a receptor site, dendrites,

and a transmission (or release) site—the axon, or neurite.

Nidicolous
A helpless chick that depends on its parents' care after birth.

Nidifugous
A chick that can move and leave the nest as soon as it breaks its shell. In less than a day, such chicks can move agilely.

Nocturnal
Active at night. Many birds of prey, such as owls, specialize in nocturnal hunting.

Nostril
Each of the openings of the nasal cavities that lead to the outside of the body.

Nutrient
Any substance obtained through diet that participates in the vital functions of a living being.

Oceanic
Region of open water beyond the edge of the continental shelf or island coasts.

Omasum
A ruminant's third stomach chamber. It is a small organ with a high absorptive capacity. It permits the recycling of water and minerals, such as sodium and phosphorus, which may return to the rumen through the saliva.

Ommatidium
The simple visual unit of a compound eye in arthropods; it contains light-sensitive cells and a lens that can form an image.

Omnivore
Animal that feeds on animal and plant species.

Operculum
Gill cover of bony fish.

Opisthoglyph
Group of snakes with fangs located in the back of the upper jaw and smaller teeth in front. The fangs can be smooth or have a groove on the surface that enables secretions to flow into the wound they produced.

Order
Taxonomic category that includes families; category lower than a class and higher than a family.

Organ
Body part made of various tissues grouped into a structural and functional unit.

Organism
Any living creature, whether single-celled or multicellular.

Osteichthyes
Class of fish that includes all bony fish, characterized by a highly ossified skeleton. This is contrasted with the class Chondrichthyes, including fish with cartilaginous skeletons (rays, skates, chimaeras, and sharks).

Ovary
The organ that produces eggs (female sex cells).

Oviduct
The duct through which the ova leave the ovary to be fertilized.

Oviparous
Refers to animals that lay eggs outside the mother's body, where they complete their development before hatching.

Ovoviviparous
Animal that reproduces by forming eggs that are carried, with soft shells, inside the female until they hatch. They may hatch inside the mother and come out as if they had been born live or be expelled from the egg pouch, breaking its membrane in order to hatch.

Papilla
Each of the small, conical elevations on skin or mucous membranes, especially those on the tongue, by means of which the sense of taste functions.

Parasite
Organism that lives at the expense of another and typically obtains nutrients that have already been processed by the host.

Parthenogenesis
Form of asexual reproduction in certain species, such as the gecko, in which the females produce young (all or mostly females) without the intervention of a male.

Patagium
The very fine membrane that joins the fingers and anterior limbs with the body, feet, and tail of bats.

Pectoral Fin
Paired fins located in the thoracic region, behind the gill openings.

Pelagic
Relating to organisms that live at or near the ocean's surface.

Pheromones
Chemical substances secreted by the reproductive glands of certain animals in order to attract individuals of the opposite sex.

Photophore
Mucous glands modified for the production of light. The light can come from symbiotic phosphorescent bacteria or from oxidation processes within the tissues.

Phylum
Taxonomic category that includes classes; category lower than a kingdom and higher than a class.

Phytoplankton
Microscopic plants, of great importance as the basic link in most underwater food chains.

Pigment
Substance that colors the skin, feathers, or tissues of animals and plants.

Placenta
The spongy tissue that completely surrounds the embryo and whose function is to allow the exchange of substances through the blood. It also protects the fetus from infections and controls physiological processes during gestation and birth.

Placoid
Scales typical of cartilaginous fish and other ancient species. These scales are made of pulp, dentine, and enamel like that found in teeth, and they have a small protrusion. They are usually very small and point outward.

Plankton
Group of small living beings, whether plants (phytoplankton) or animals (zooplankton), that live suspended in freshwater or ocean water.

Plantigrade
Refers to mammals that use the entire foot in walking. Humans are plantigrade.

Planula
Type of unattached, ciliated larva of

many organisms of the phylum Cnidaria (jellyfish, sea anemones, and coral).

Plastron
Lower part of the shell of a turtle or tortoise.

Polyandry
Refers to the relationship in which a female copulates with various males during one breeding period.

Polygamy
Reproductive relationship between one animal of one sex and several animals of the other. When one male mates with several females, it is called polygyny. Only rarely do females have multiple male reproductive partners (polyandry).

Polygyny
The social system of certain animals, in which the male gathers a harem of females.

Polyp
The immobile stage in the life cycle of animals of the phylum Cnidaria.

Population
Group of individuals of the same species that live in a certain area during a specific time.

Predator
Organism that feeds on other living beings.

Prey
Animal hunted by another to serve as food for the latter. Animals that hunt prey are called predators.

Protein
Macromolecule composed of one or more chains of amino acids. They define the physical characteristics of an organism and, when acting as enzymes, regulate chemical reactions.

Protein Cord
Embryonic structure: each of the two filaments that sustain the yolk of the egg within the white.

Proteroglyph
System of fangs in cobras, mambas, coral snakes, and sea snakes; or the name referring to the group that contains these types of snakes. The fangs are located in the front of the upper jaw and are hollow or have a surface groove for carrying venom. They are relatively short and are fixed in an extended position.

Protractile
Describes a type of reptilian tongue that can be voluntarily hurled outward in an extremely rapid, precise movement.

Proventriculus
The first portion of the stomach, or the true stomach, of a bird. The other portion of a bird's stomach is the gizzard.

Quadruped
Refers to a four-legged animal.

Radial Symmetry
The regular disposition of body parts around a central axis in such a way that any plane that cuts through the axis divides the organism in halves that constitute mirror images of each other. It is seen in adult echinoderms.

Rectrices
Technical term used by ornithologists to describe a bird's tail feathers.

Reef
Hard bank that barely reaches above the ocean surface or that lies in very shallow waters. It can pose a danger for navigation. It can be inorganic in nature or result from the growth of coral.

Reflex
Simple action of the nervous system that involves a sensory neuron, often one or more interneurons, and one or more motor neurons.

Reticulum
The second chamber of a ruminant's stomach. It is a crossroad where the particles that enter and leave the rumen are separated. Only small particles of less than $1/12$ of an inch (2 mm) or dense ones greater than 1 ounce per inch (1.2 g per mm) can go on to the third chamber.

Retina
The inner membrane of the eyes of mammals and other animals, where light sensations are transformed into nerve impulses.

Rod
Along with cones, rods form the photoreceptor cells of the retina of vertebrates. They are responsible for peripheral and night vision, though they perceive colors poorly.

Rumen
The first chamber of a ruminant's

stomach. It is a large fermentation vessel that can hold up to 220–265 pounds (100–120 kg) of matter in the process of being digested. Fiber particles remain there between 20 and 48 hours.

Ruminate
The process of chewing food a second time, returning food to the mouth that was already in the chamber that certain animals (ruminants) have.

Salinity
Measurement of the amount of common salt in water or soil. Common salt is a sodium salt, sodium chloride, common in nature, that gives a salty flavor to ocean water and salt lakes.

Sarcopterygii
Another name for the Choanichthyes, a subclass of bony fish. Their fins are joined to the body by fleshy lobes, and those of the lungfish resemble filaments.

Scales
Small bony plates that grow from the skin and overlap each other.

Scapula
Triangular bone, also called the shoulder blade. With the clavicle, it forms the scapular belt.

Scavenger
Animals that eat organic forms of life that have died. They help maintain the equilibrium of the ecosystem by feeding upon dead animals, breaking them down.

School
Transient grouping of fish of the same population or species, brought together by similar behavior.

Segmentation
Successive cell divisions in the egg of an animal to form a multicellular blastula.

Simple Metamorphosis
Process in which the general appearance of an animal remains similar, although some organs atrophy and others develop.

Solenoglyph
System of long, hollow fangs in some snakes or the name of the group that refers to snakes possessing this characteristic. The fangs are the only teeth in the upper jaw, and they pivot so that they lie flat along the roof of the mouth when the mouth is closed. They inject venom deep into the tissues of prey.

Song
Sound or series of sounds produced by a bird to demarcate its territory or to find a mate. The songs of birds can be simple or elaborate, and some are very melodic.

Songbirds
Singing birds. Passerines include songbirds.

Spawning
Action of producing or laying eggs.

Species
A group of individuals that recognize one another as belonging to the same reproductive unit.

Spermaceti
A waxy substance contained in the organ that bears the same name, located in the head of the sperm whale. It is believed that it aids deep dives, although some specialists believe that it may assist echolocation.

Spinal Cord
An extension of the central nervous system. Often protected by vertebrae, this soft, fatty material is the major nerve pathway that carries information to and from the brain and muscles.

Spiracle
One of the external openings of the respiratory system in terrestrial arthropods.

Stinger
Sharp point that grows from the skin. The order Rajiformes includes two families that have poisonous stingers on the final one-third of their tail. The stinger is extremely sharp and has serrated edges.

Striated Muscle
Muscle tissue with a striped appearance that shows the arrangement of the contracting elements. Includes the voluntary skeletal muscle and the cardiac muscle.

Substrate
The surface that constitutes an organism's habitat or life support.

Sucker
Structure formed from the pectoral and pelvic fins to generate pressure and stick to a surface. It can also be a modification of the anterior dorsal fin, the pelvic fin, or the buccal (mouth) disk of the cyclostomes.

Swamp
Depression on the ground in which water is gathered, sometimes called a marsh. Its bottom is more or less boggy. It is the habitat of many wading birds.

Swarm
Insects that act in a group for eating, mating, or finding a new location for a nest.

Swim Bladder
A sac located in the anterior dorsal region of the intestine that contains gas. Its function is to enable the animal to maintain buoyancy. This structure evolved as a lung, and, in some fish, it retains its breathing function.

Tentacles
Long and flexible organs located around the mouth of many invertebrates, often prehensile and tactile.

Thermal
Hot air current that rises. Many birds make use of it to gain height effortlessly.

Thorax
In crustaceans and insects, the fused segments located between the head and the abdomen to which the legs are attached.

Tissue
Group of identical cells that carry out a common function.

Trachea
In insects and some other terrestrial arthropods, the system of air conduits covered with chitin.

Trophic Chain
System formed by a group of living beings that successively feed on each other.

Tundra
Vast plains without trees in the Arctic regions of northern Asia, Europe, and North America.

Ungulate
A mammal that supports itself and walks on the tips of its digits, which are covered by a hoof.

Uric Acid
Water-insoluble nitrogenated waste product; the main component of the excrement of reptiles and insects.

Uropatagium
The membrane that bats have between their feet. It also encloses the tail.

Venom
Chemical agent injected into other animals in order to kill or paralyze them, or to ward off an attack.

Ventral Fin
Paired fins located on the abdomen.

Vertebrate
Animals that have a spinal column, such as birds, fish, reptiles, amphibians, and mammals.

Viviparous
Refers to animals in which the embryonic development of offspring occurs inside the mother's body and the offspring emerge as viable young at birth.

Warren
A burrow where certain animals raise their young.

Weaning
The process by which a mammal ceases to receive maternal milk as its subsistence.

Yolk
Yellow part of the egg. If the egg is fertilized, a small embryo grows that will use the yolk (and white) as food.

Zoonosis
Illness transmitted by animals to humans.

Zooplankton
Microscopic larvae of crustaceans, fish, and other sea animals.

INDEX